Observational Astronomy: A Very Short Introduction

VERY SHORT INTRODUCTIONS are for anyone wanting a stimulating and accessible way into a new subject. They are written by experts, and have been translated into more than 45 different languages.

The series began in 1995, and now covers a wide variety of topics in every discipline. The VSI library currently contains over 700 volumes—a Very Short Introduction to everything from Psychology and Philosophy of Science to American History and Relativity—and continues to grow in every subject area.

Very Short Introductions available now:

ABOLITIONISM Richard S. Newman
THE ABRAHAMIC RELIGIONS Charles L. Cohen
ACCOUNTING Christopher Nobes
ADDICTION Keith Humphreys
ADOLESCENCE Peter K. Smith
THEODOR W. ADORNO Andrew Bowie
ADVERTISING Winston Fletcher
AERIAL WARFARE Frank Ledwidge
AESTHETICS Bence Nanay
AFRICAN AMERICAN HISTORY Jonathan Scott Holloway
AFRICAN AMERICAN RELIGION Eddie S. Glaude Jr
AFRICAN HISTORY John Parker and Richard Rathbone
AFRICAN POLITICS Ian Taylor
AFRICAN RELIGIONS Jacob K. Olupona
AGEING Nancy A. Pachana
AGNOSTICISM Robin Le Poidevin
AGRICULTURE Paul Brassley and Richard Soffe
ALEXANDER THE GREAT Hugh Bowden
ALGEBRA Peter M. Higgins
AMERICAN BUSINESS HISTORY Walter A. Friedman
AMERICAN CULTURAL HISTORY Eric Avila
AMERICAN FOREIGN RELATIONS Andrew Preston
AMERICAN HISTORY Paul S. Boyer
AMERICAN IMMIGRATION David A. Gerber
AMERICAN INTELLECTUAL HISTORY Jennifer Ratner-Rosenhagen
THE AMERICAN JUDICIAL SYSTEM Charles L. Zelden
AMERICAN LEGAL HISTORY G. Edward White
AMERICAN MILITARY HISTORY Joseph T. Glatthaar
AMERICAN NAVAL HISTORY Craig L. Symonds
AMERICAN POETRY David Caplan
AMERICAN POLITICAL HISTORY Donald Critchlow
AMERICAN POLITICAL PARTIES AND ELECTIONS L. Sandy Maisel
AMERICAN POLITICS Richard M. Valelly
THE AMERICAN PRESIDENCY Charles O. Jones
THE AMERICAN REVOLUTION Robert J. Allison
AMERICAN SLAVERY Heather Andrea Williams
THE AMERICAN SOUTH Charles Reagan Wilson
THE AMERICAN WEST Stephen Aron
AMERICAN WOMEN'S HISTORY Susan Ware
AMPHIBIANS T. S. Kemp
ANAESTHESIA Aidan O'Donnell

ANALYTIC PHILOSOPHY Michael Beaney
ANARCHISM Alex Prichard
ANCIENT ASSYRIA Karen Radner
ANCIENT EGYPT Ian Shaw
ANCIENT EGYPTIAN ART AND ARCHITECTURE Christina Riggs
ANCIENT GREECE Paul Cartledge
ANCIENT GREEK AND ROMAN SCIENCE Liba Taub
THE ANCIENT NEAR EAST Amanda H. Podany
ANCIENT PHILOSOPHY Julia Annas
ANCIENT WARFARE Harry Sidebottom
ANGELS David Albert Jones
ANGLICANISM Mark Chapman
THE ANGLO-SAXON AGE John Blair
ANIMAL BEHAVIOUR Tristram D. Wyatt
THE ANIMAL KINGDOM Peter Holland
ANIMAL RIGHTS David DeGrazia
ANSELM Thomas Williams
THE ANTARCTIC Klaus Dodds
ANTHROPOCENE Erle C. Ellis
ANTISEMITISM Steven Beller
ANXIETY Daniel Freeman and Jason Freeman
THE APOCRYPHAL GOSPELS Paul Foster
APPLIED MATHEMATICS Alain Goriely
THOMAS AQUINAS Fergus Kerr
ARBITRATION Thomas Schultz and Thomas Grant
ARCHAEOLOGY Paul Bahn
ARCHITECTURE Andrew Ballantyne
THE ARCTIC Klaus Dodds and Jamie Woodward
HANNAH ARENDT Dana Villa
ARISTOCRACY William Doyle
ARISTOTLE Jonathan Barnes
ART HISTORY Dana Arnold
ART THEORY Cynthia Freeland
ARTIFICIAL INTELLIGENCE Margaret A. Boden
ASIAN AMERICAN HISTORY Madeline Y. Hsu
ASTROBIOLOGY David C. Catling
ASTROPHYSICS James Binney
ATHEISM Julian Baggini
THE ATMOSPHERE Paul I. Palmer
AUGUSTINE Henry Chadwick
JANE AUSTEN Tom Keymer
AUSTRALIA Kenneth Morgan
AUTISM Uta Frith
AUTOBIOGRAPHY Laura Marcus
THE AVANT GARDE David Cottington
THE AZTECS Davíd Carrasco
BABYLONIA Trevor Bryce
BACTERIA Sebastian G. B. Amyes
BANKING John Goddard and John O. S. Wilson
BARTHES Jonathan Culler
THE BEATS David Sterritt
BEAUTY Roger Scruton
LUDWIG VAN BEETHOVEN Mark Evan Bonds
BEHAVIOURAL ECONOMICS Michelle Baddeley
BESTSELLERS John Sutherland
THE BIBLE John Riches
BIBLICAL ARCHAEOLOGY Eric H. Cline
BIG DATA Dawn E. Holmes
BIOCHEMISTRY Mark Lorch
BIOGEOGRAPHY Mark V. Lomolino
BIOGRAPHY Hermione Lee
BIOMETRICS Michael Fairhurst
ELIZABETH BISHOP Jonathan F. S. Post
BLACK HOLES Katherine Blundell
BLASPHEMY Yvonne Sherwood
BLOOD Chris Cooper
THE BLUES Elijah Wald
THE BODY Chris Shilling
THE BOHEMIANS David Weir
NIELS BOHR J. L. Heilbron
THE BOOK OF COMMON PRAYER Brian Cummings
THE BOOK OF MORMON Terryl Givens
BORDERS Alexander C. Diener and Joshua Hagen
THE BRAIN Michael O'Shea
BRANDING Robert Jones
THE BRICS Andrew F. Cooper
BRITISH CINEMA Charles Barr

THE BRITISH CONSTITUTION
Martin Loughlin
THE BRITISH EMPIRE Ashley Jackson
BRITISH POLITICS Tony Wright
BUDDHA Michael Carrithers
BUDDHISM Damien Keown
BUDDHIST ETHICS Damien Keown
BYZANTIUM Peter Sarris
CALVINISM Jon Balserak
ALBERT CAMUS Oliver Gloag
CANADA Donald Wright
CANCER Nicholas James
CAPITALISM James Fulcher
CATHOLICISM Gerald O'Collins
CAUSATION Stephen Mumford and
Rani Lill Anjum
THE CELL Terence Allen and
Graham Cowling
THE CELTS Barry Cunliffe
CHAOS Leonard Smith
GEOFFREY CHAUCER David Wallace
CHEMISTRY Peter Atkins
CHILD PSYCHOLOGY Usha Goswami
CHILDREN'S LITERATURE
Kimberley Reynolds
CHINESE LITERATURE Sabina Knight
CHOICE THEORY Michael Allingham
CHRISTIAN ART Beth Williamson
CHRISTIAN ETHICS D. Stephen Long
CHRISTIANITY Linda Woodhead
CIRCADIAN RHYTHMS
Russell Foster and Leon Kreitzman
CITIZENSHIP Richard Bellamy
CITY PLANNING Carl Abbott
CIVIL ENGINEERING
David Muir Wood
THE CIVIL RIGHTS MOVEMENT
Thomas C. Holt
CLASSICAL LITERATURE
William Allan
CLASSICAL MYTHOLOGY
Helen Morales
CLASSICS Mary Beard and
John Henderson
CLAUSEWITZ Michael Howard
CLIMATE Mark Maslin
CLIMATE CHANGE Mark Maslin
CLINICAL PSYCHOLOGY
Susan Llewelyn and
Katie Aafjes-van Doorn
COGNITIVE BEHAVIOURAL
THERAPY Freda McManus
COGNITIVE NEUROSCIENCE
Richard Passingham
THE COLD WAR Robert J. McMahon
COLONIAL AMERICA Alan Taylor
COLONIAL LATIN AMERICAN
LITERATURE Rolena Adorno
COMBINATORICS Robin Wilson
COMEDY Matthew Bevis
COMMUNISM Leslie Holmes
COMPARATIVE LITERATURE
Ben Hutchinson
COMPETITION AND ANTITRUST
LAW Ariel Ezrachi
COMPLEXITY John H. Holland
THE COMPUTER Darrel Ince
COMPUTER SCIENCE
Subrata Dasgupta
CONCENTRATION CAMPS
Dan Stone
CONDENSED MATTER PHYSICS
Ross H. McKenzie
CONFUCIANISM Daniel K. Gardner
THE CONQUISTADORS
Matthew Restall and
Felipe Fernández-Armesto
CONSCIENCE Paul Strohm
CONSCIOUSNESS Susan Blackmore
CONTEMPORARY ART
Julian Stallabrass
CONTEMPORARY FICTION
Robert Eaglestone
CONTINENTAL PHILOSOPHY
Simon Critchley
COPERNICUS Owen Gingerich
CORAL REEFS Charles Sheppard
CORPORATE SOCIAL
RESPONSIBILITY Jeremy Moon
CORRUPTION Leslie Holmes
COSMOLOGY Peter Coles
COUNTRY MUSIC Richard Carlin
CREATIVITY Vlad Glăveanu
CRIME FICTION Richard Bradford
CRIMINAL JUSTICE
Julian V. Roberts
CRIMINOLOGY Tim Newburn
CRITICAL THEORY
Stephen Eric Bronner
THE CRUSADES Christopher Tyerman

CRYPTOGRAPHY Fred Piper and Sean Murphy
CRYSTALLOGRAPHY A. M. Glazer
THE CULTURAL REVOLUTION Richard Curt Kraus
DADA AND SURREALISM David Hopkins
DANTE Peter Hainsworth and David Robey
DARWIN Jonathan Howard
THE DEAD SEA SCROLLS Timothy H. Lim
DECADENCE David Weir
DECOLONIZATION Dane Kennedy
DEMENTIA Kathleen Taylor
DEMOCRACY Bernard Crick
DEMOGRAPHY Sarah Harper
DEPRESSION Jan Scott and Mary Jane Tacchi
DERRIDA Simon Glendinning
DESCARTES Tom Sorell
DESERTS Nick Middleton
DESIGN John Heskett
DEVELOPMENT Ian Goldin
DEVELOPMENTAL BIOLOGY Lewis Wolpert
THE DEVIL Darren Oldridge
DIASPORA Kevin Kenny
CHARLES DICKENS Jenny Hartley
DICTIONARIES Lynda Mugglestone
DINOSAURS David Norman
DIPLOMATIC HISTORY Joseph M. Siracusa
DOCUMENTARY FILM Patricia Aufderheide
DREAMING J. Allan Hobson
DRUGS Les Iversen
DRUIDS Barry Cunliffe
DYNASTY Jeroen Duindam
DYSLEXIA Margaret J. Snowling
EARLY MUSIC Thomas Forrest Kelly
THE EARTH Martin Redfern
EARTH SYSTEM SCIENCE Tim Lenton
ECOLOGY Jaboury Ghazoul
ECONOMICS Partha Dasgupta
EDUCATION Gary Thomas
EGYPTIAN MYTH Geraldine Pinch
EIGHTEENTH-CENTURY BRITAIN Paul Langford
THE ELEMENTS Philip Ball
EMOTION Dylan Evans
EMPIRE Stephen Howe
EMPLOYMENT LAW David Cabrelli
ENERGY SYSTEMS Nick Jenkins
ENGELS Terrell Carver
ENGINEERING David Blockley
THE ENGLISH LANGUAGE Simon Horobin
ENGLISH LITERATURE Jonathan Bate
THE ENLIGHTENMENT John Robertson
ENTREPRENEURSHIP Paul Westhead and Mike Wright
ENVIRONMENTAL ECONOMICS Stephen Smith
ENVIRONMENTAL ETHICS Robin Attfield
ENVIRONMENTAL LAW Elizabeth Fisher
ENVIRONMENTAL POLITICS Andrew Dobson
ENZYMES Paul Engel
EPICUREANISM Catherine Wilson
EPIDEMIOLOGY Rodolfo Saracci
ETHICS Simon Blackburn
ETHNOMUSICOLOGY Timothy Rice
THE ETRUSCANS Christopher Smith
EUGENICS Philippa Levine
THE EUROPEAN UNION Simon Usherwood and John Pinder
EUROPEAN UNION LAW Anthony Arnull
EVANGELICALISM John G. Stackhouse Jr.
EVIL Luke Russell
EVOLUTION Brian and Deborah Charlesworth
EXISTENTIALISM Thomas Flynn
EXPLORATION Stewart A. Weaver
EXTINCTION Paul B. Wignall
THE EYE Michael Land
FAIRY TALE Marina Warner
FAMILY LAW Jonathan Herring
MICHAEL FARADAY Frank A. J. L. James
FASCISM Kevin Passmore
FASHION Rebecca Arnold
FEDERALISM Mark J. Rozell and Clyde Wilcox
FEMINISM Margaret Walters

FILM Michael Wood
FILM MUSIC Kathryn Kalinak
FILM NOIR James Naremore
FIRE Andrew C. Scott
THE FIRST WORLD WAR Michael Howard
FLUID MECHANICS Eric Lauga
FOLK MUSIC Mark Slobin
FOOD John Krebs
FORENSIC PSYCHOLOGY David Canter
FORENSIC SCIENCE Jim Fraser
FORESTS Jaboury Ghazoul
FOSSILS Keith Thomson
FOUCAULT Gary Gutting
THE FOUNDING FATHERS R. B. Bernstein
FRACTALS Kenneth Falconer
FREE SPEECH Nigel Warburton
FREE WILL Thomas Pink
FREEMASONRY Andreas Önnerfors
FRENCH LITERATURE John D. Lyons
FRENCH PHILOSOPHY Stephen Gaukroger and Knox Peden
THE FRENCH REVOLUTION William Doyle
FREUD Anthony Storr
FUNDAMENTALISM Malise Ruthven
FUNGI Nicholas P. Money
THE FUTURE Jennifer M. Gidley
GALAXIES John Gribbin
GALILEO Stillman Drake
GAME THEORY Ken Binmore
GANDHI Bhikhu Parekh
GARDEN HISTORY Gordon Campbell
GENES Jonathan Slack
GENIUS Andrew Robinson
GENOMICS John Archibald
GEOGRAPHY John Matthews and David Herbert
GEOLOGY Jan Zalasiewicz
GEOMETRY Maciej Dunajski
GEOPHYSICS William Lowrie
GEOPOLITICS Klaus Dodds
GERMAN LITERATURE Nicholas Boyle
GERMAN PHILOSOPHY Andrew Bowie
THE GHETTO Bryan Cheyette
GLACIATION David J. A. Evans
GLOBAL CATASTROPHES Bill McGuire
GLOBAL ECONOMIC HISTORY Robert C. Allen
GLOBAL ISLAM Nile Green
GLOBALIZATION Manfred B. Steger
GOD John Bowker
GÖDEL'S THEOREM A. W. Moore
GOETHE Ritchie Robertson
THE GOTHIC Nick Groom
GOVERNANCE Mark Bevir
GRAVITY Timothy Clifton
THE GREAT DEPRESSION AND THE NEW DEAL Eric Rauchway
HABEAS CORPUS Amanda L. Tyler
HABERMAS James Gordon Finlayson
THE HABSBURG EMPIRE Martyn Rady
HAPPINESS Daniel M. Haybron
THE HARLEM RENAISSANCE Cheryl A. Wall
THE HEBREW BIBLE AS LITERATURE Tod Linafelt
HEGEL Peter Singer
HEIDEGGER Michael Inwood
THE HELLENISTIC AGE Peter Thonemann
HEREDITY John Waller
HERMENEUTICS Jens Zimmermann
HERODOTUS Jennifer T. Roberts
HIEROGLYPHS Penelope Wilson
HINDUISM Kim Knott
HISTORY John H. Arnold
THE HISTORY OF ASTRONOMY Michael Hoskin
THE HISTORY OF CHEMISTRY William H. Brock
THE HISTORY OF CHILDHOOD James Marten
THE HISTORY OF CINEMA Geoffrey Nowell-Smith
THE HISTORY OF COMPUTING Doron Swade
THE HISTORY OF EMOTIONS Thomas Dixon
THE HISTORY OF LIFE Michael Benton
THE HISTORY OF MATHEMATICS Jacqueline Stedall
THE HISTORY OF MEDICINE William Bynum
THE HISTORY OF PHYSICS J. L. Heilbron

THE HISTORY OF POLITICAL THOUGHT Richard Whatmore
THE HISTORY OF TIME Leofranc Holford-Strevens
HIV AND AIDS Alan Whiteside
HOBBES Richard Tuck
HOLLYWOOD Peter Decherney
THE HOLY ROMAN EMPIRE Joachim Whaley
HOME Michael Allen Fox
HOMER Barbara Graziosi
HORMONES Martin Luck
HORROR Darryl Jones
HUMAN ANATOMY Leslie Klenerman
HUMAN EVOLUTION Bernard Wood
HUMAN PHYSIOLOGY Jamie A. Davies
HUMAN RESOURCE MANAGEMENT Adrian Wilkinson
HUMAN RIGHTS Andrew Clapham
HUMANISM Stephen Law
HUME James A. Harris
HUMOUR Noël Carroll
THE ICE AGE Jamie Woodward
IDENTITY Florian Coulmas
IDEOLOGY Michael Freeden
THE IMMUNE SYSTEM Paul Klenerman
INDIAN CINEMA Ashish Rajadhyaksha
INDIAN PHILOSOPHY Sue Hamilton
THE INDUSTRIAL REVOLUTION Robert C. Allen
INFECTIOUS DISEASE Marta L. Wayne and Benjamin M. Bolker
INFINITY Ian Stewart
INFORMATION Luciano Floridi
INNOVATION Mark Dodgson and David Gann
INTELLECTUAL PROPERTY Siva Vaidhyanathan
INTELLIGENCE Ian J. Deary
INTERNATIONAL LAW Vaughan Lowe
INTERNATIONAL MIGRATION Khalid Koser
INTERNATIONAL RELATIONS Christian Reus-Smit
INTERNATIONAL SECURITY Christopher S. Browning
INSECTS Simon Leather
IRAN Ali M. Ansari
ISLAM Malise Ruthven
ISLAMIC HISTORY Adam Silverstein
ISLAMIC LAW Mashood A. Baderin
ISOTOPES Rob Ellam
ITALIAN LITERATURE Peter Hainsworth and David Robey
HENRY JAMES Susan L. Mizruchi
JAPANESE LITERATURE Alan Tansman
JESUS Richard Bauckham
JEWISH HISTORY David N. Myers
JEWISH LITERATURE Ilan Stavans
JOURNALISM Ian Hargreaves
JAMES JOYCE Colin MacCabe
JUDAISM Norman Solomon
JUNG Anthony Stevens
THE JURY Renée Lettow Lerner
KABBALAH Joseph Dan
KAFKA Ritchie Robertson
KANT Roger Scruton
KEYNES Robert Skidelsky
KIERKEGAARD Patrick Gardiner
KNOWLEDGE Jennifer Nagel
THE KORAN Michael Cook
KOREA Michael J. Seth
LAKES Warwick F. Vincent
LANDSCAPE ARCHITECTURE Ian H. Thompson
LANDSCAPES AND GEOMORPHOLOGY Andrew Goudie and Heather Viles
LANGUAGES Stephen R. Anderson
LATE ANTIQUITY Gillian Clark
LAW Raymond Wacks
THE LAWS OF THERMODYNAMICS Peter Atkins
LEADERSHIP Keith Grint
LEARNING Mark Haselgrove
LEIBNIZ Maria Rosa Antognazza
C. S. LEWIS James Como
LIBERALISM Michael Freeden
LIGHT Ian Walmsley
LINCOLN Allen C. Guelzo
LINGUISTICS Peter Matthews
LITERARY THEORY Jonathan Culler
LOCKE John Dunn
LOGIC Graham Priest

LOVE Ronald de Sousa
MARTIN LUTHER Scott H. Hendrix
MACHIAVELLI Quentin Skinner
MADNESS Andrew Scull
MAGIC Owen Davies
MAGNA CARTA Nicholas Vincent
MAGNETISM Stephen Blundell
MALTHUS Donald Winch
MAMMALS T. S. Kemp
MANAGEMENT John Hendry
NELSON MANDELA Elleke Boehmer
MAO Delia Davin
MARINE BIOLOGY Philip V. Mladenov
MARKETING Kenneth Le Meunier-FitzHugh
THE MARQUIS DE SADE John Phillips
MARTYRDOM Jolyon Mitchell
MARX Peter Singer
MATERIALS Christopher Hall
MATHEMATICAL ANALYSIS Richard Earl
MATHEMATICAL FINANCE Mark H. A. Davis
MATHEMATICS Timothy Gowers
MATTER Geoff Cottrell
THE MAYA Matthew Restall and Amara Solari
THE MEANING OF LIFE Terry Eagleton
MEASUREMENT David Hand
MEDICAL ETHICS Michael Dunn and Tony Hope
MEDICAL LAW Charles Foster
MEDIEVAL BRITAIN John Gillingham and Ralph A. Griffiths
MEDIEVAL LITERATURE Elaine Treharne
MEDIEVAL PHILOSOPHY John Marenbon
MEMORY Jonathan K. Foster
METAPHYSICS Stephen Mumford
METHODISM William J. Abraham
THE MEXICAN REVOLUTION Alan Knight
MICROBIOLOGY Nicholas P. Money
MICROBIOMES Angela E. Douglas
MICROECONOMICS Avinash Dixit
MICROSCOPY Terence Allen
THE MIDDLE AGES Miri Rubin
MILITARY JUSTICE Eugene R. Fidell
MILITARY STRATEGY Antulio J. Echevarria II
JOHN STUART MILL Gregory Claeys
MINERALS David Vaughan
MIRACLES Yujin Nagasawa
MODERN ARCHITECTURE Adam Sharr
MODERN ART David Cottington
MODERN BRAZIL Anthony W. Pereira
MODERN CHINA Rana Mitter
MODERN DRAMA Kirsten E. Shepherd-Barr
MODERN FRANCE Vanessa R. Schwartz
MODERN INDIA Craig Jeffrey
MODERN IRELAND Senia Pašeta
MODERN ITALY Anna Cento Bull
MODERN JAPAN Christopher Goto-Jones
MODERN LATIN AMERICAN LITERATURE Roberto González Echevarría
MODERN WAR Richard English
MODERNISM Christopher Butler
MOLECULAR BIOLOGY Aysha Divan and Janice A. Royds
MOLECULES Philip Ball
MONASTICISM Stephen J. Davis
THE MONGOLS Morris Rossabi
MONTAIGNE William M. Hamlin
MOONS David A. Rothery
MORMONISM Richard Lyman Bushman
MOUNTAINS Martin F. Price
MUHAMMAD Jonathan A. C. Brown
MULTICULTURALISM Ali Rattansi
MULTILINGUALISM John C. Maher
MUSIC Nicholas Cook
MUSIC AND TECHNOLOGY Mark Katz
MYTH Robert A. Segal
NANOTECHNOLOGY Philip Moriarty
NAPOLEON David A. Bell
THE NAPOLEONIC WARS Mike Rapport
NATIONALISM Steven Grosby
NATIVE AMERICAN LITERATURE Sean Teuton
NAVIGATION Jim Bennett
NAZI GERMANY Jane Caplan

NEGOTIATION Carrie Menkel-Meadow
NEOLIBERALISM Manfred B. Steger and Ravi K. Roy
NETWORKS Guido Caldarelli and Michele Catanzaro
THE NEW TESTAMENT Luke Timothy Johnson
THE NEW TESTAMENT AS LITERATURE Kyle Keefer
NEWTON Robert Iliffe
NIETZSCHE Michael Tanner
NINETEENTH-CENTURY BRITAIN Christopher Harvie and H. C. G. Matthew
THE NORMAN CONQUEST George Garnett
NORTH AMERICAN INDIANS Theda Perdue and Michael D. Green
NORTHERN IRELAND Marc Mulholland
NOTHING Frank Close
NUCLEAR PHYSICS Frank Close
NUCLEAR POWER Maxwell Irvine
NUCLEAR WEAPONS Joseph M. Siracusa
NUMBER THEORY Robin Wilson
NUMBERS Peter M. Higgins
NUTRITION David A. Bender
OBJECTIVITY Stephen Gaukroger
OBSERVATIONAL ASTRONOMY Geoff Cottrell
OCEANS Dorrik Stow
THE OLD TESTAMENT Michael D. Coogan
THE ORCHESTRA D. Kern Holoman
ORGANIC CHEMISTRY Graham Patrick
ORGANIZATIONS Mary Jo Hatch
ORGANIZED CRIME Georgios A. Antonopoulos and Georgios Papanicolaou
ORTHODOX CHRISTIANITY A. Edward Siecienski
OVID Llewelyn Morgan
PAGANISM Owen Davies
PAKISTAN Pippa Virdee
THE PALESTINIAN-ISRAELI CONFLICT Martin Bunton
PANDEMICS Christian W. McMillen
PARTICLE PHYSICS Frank Close
PAUL E. P. Sanders
IVAN PAVLOV Daniel P. Todes
PEACE Oliver P. Richmond
PENTECOSTALISM William K. Kay
PERCEPTION Brian Rogers
THE PERIODIC TABLE Eric R. Scerri
PHILOSOPHICAL METHOD Timothy Williamson
PHILOSOPHY Edward Craig
PHILOSOPHY IN THE ISLAMIC WORLD Peter Adamson
PHILOSOPHY OF BIOLOGY Samir Okasha
PHILOSOPHY OF LAW Raymond Wacks
PHILOSOPHY OF MIND Barbara Gail Montero
PHILOSOPHY OF PHYSICS David Wallace
PHILOSOPHY OF SCIENCE Samir Okasha
PHILOSOPHY OF RELIGION Tim Bayne
PHOTOGRAPHY Steve Edwards
PHYSICAL CHEMISTRY Peter Atkins
PHYSICS Sidney Perkowitz
PILGRIMAGE Ian Reader
PLAGUE Paul Slack
PLANETARY SYSTEMS Raymond T. Pierrehumbert
PLANETS David A. Rothery
PLANTS Timothy Walker
PLATE TECTONICS Peter Molnar
PLATO Julia Annas
POETRY Bernard O'Donoghue
POLITICAL PHILOSOPHY David Miller
POLITICS Kenneth Minogue
POLYGAMY Sarah M. S. Pearsall
POPULISM Cas Mudde and Cristóbal Rovira Kaltwasser
POSTCOLONIALISM Robert J. C. Young
POSTMODERNISM Christopher Butler
POSTSTRUCTURALISM Catherine Belsey
POVERTY Philip N. Jefferson
PREHISTORY Chris Gosden
PRESOCRATIC PHILOSOPHY Catherine Osborne
PRIVACY Raymond Wacks

PROBABILITY John Haigh
PROGRESSIVISM Walter Nugent
PROHIBITION W. J. Rorabaugh
PROJECTS Andrew Davies
PROTESTANTISM Mark A. Noll
PSEUDOSCIENCE Michael D. Gordin.
PSYCHIATRY Tom Burns
PSYCHOANALYSIS Daniel Pick
PSYCHOLOGY Gillian Butler and Freda McManus
PSYCHOLOGY OF MUSIC Elizabeth Hellmuth Margulis
PSYCHOPATHY Essi Viding
PSYCHOTHERAPY Tom Burns and Eva Burns-Lundgren
PUBLIC ADMINISTRATION Stella Z. Theodoulou and Ravi K. Roy
PUBLIC HEALTH Virginia Berridge
PURITANISM Francis J. Bremer
THE QUAKERS Pink Dandelion
QUANTUM THEORY John Polkinghorne
RACISM Ali Rattansi
RADIOACTIVITY Claudio Tuniz
RASTAFARI Ennis B. Edmonds
READING Belinda Jack
THE REAGAN REVOLUTION Gil Troy
REALITY Jan Westerhoff
RECONSTRUCTION Allen C. Guelzo
THE REFORMATION Peter Marshall
REFUGEES Gil Loescher
RELATIVITY Russell Stannard
RELIGION Thomas A. Tweed
RELIGION IN AMERICA Timothy Beal
THE RENAISSANCE Jerry Brotton
RENAISSANCE ART Geraldine A. Johnson
RENEWABLE ENERGY Nick Jelley
REPTILES T. S. Kemp
REVOLUTIONS Jack A. Goldstone
RHETORIC Richard Toye
RISK Baruch Fischhoff and John Kadvany
RITUAL Barry Stephenson
RIVERS Nick Middleton
ROBOTICS Alan Winfield
ROCKS Jan Zalasiewicz
ROMAN BRITAIN Peter Salway
THE ROMAN EMPIRE Christopher Kelly
THE ROMAN REPUBLIC David M. Gwynn
ROMANTICISM Michael Ferber
ROUSSEAU Robert Wokler
RUSSELL A. C. Grayling
THE RUSSIAN ECONOMY Richard Connolly
RUSSIAN HISTORY Geoffrey Hosking
RUSSIAN LITERATURE Catriona Kelly
THE RUSSIAN REVOLUTION S. A. Smith
SAINTS Simon Yarrow
SAMURAI Michael Wert
SAVANNAS Peter A. Furley
SCEPTICISM Duncan Pritchard
SCHIZOPHRENIA Chris Frith and Eve Johnstone
SCHOPENHAUER Christopher Janaway
SCIENCE AND RELIGION Thomas Dixon and Adam R. Shapiro
SCIENCE FICTION David Seed
THE SCIENTIFIC REVOLUTION Lawrence M. Principe
SCOTLAND Rab Houston
SECULARISM Andrew Copson
SEXUAL SELECTION Marlene Zuk and Leigh W. Simmons
SEXUALITY Véronique Mottier
WILLIAM SHAKESPEARE Stanley Wells
SHAKESPEARE'S COMEDIES Bart van Es
SHAKESPEARE'S SONNETS AND POEMS Jonathan F. S. Post
SHAKESPEARE'S TRAGEDIES Stanley Wells
GEORGE BERNARD SHAW Christopher Wixson
MARY SHELLEY Charlotte Gordon
THE SHORT STORY Andrew Kahn
SIKHISM Eleanor Nesbitt
SILENT FILM Donna Kornhaber
THE SILK ROAD James A. Millward
SLANG Jonathon Green
SLEEP Steven W. Lockley and Russell G. Foster
SMELL Matthew Cobb
ADAM SMITH Christopher J. Berry
SOCIAL AND CULTURAL ANTHROPOLOGY John Monaghan and Peter Just

SOCIAL PSYCHOLOGY Richard J. Crisp
SOCIAL WORK Sally Holland and Jonathan Scourfield
SOCIALISM Michael Newman
SOCIOLINGUISTICS John Edwards
SOCIOLOGY Steve Bruce
SOCRATES C. C. W. Taylor
SOFT MATTER Tom McLeish
SOUND Mike Goldsmith
SOUTHEAST ASIA James R. Rush
THE SOVIET UNION Stephen Lovell
THE SPANISH CIVIL WAR Helen Graham
SPANISH LITERATURE Jo Labanyi
THE SPARTANS Andrew J. Bayliss
SPINOZA Roger Scruton
SPIRITUALITY Philip Sheldrake
SPORT Mike Cronin
STARS Andrew King
STATISTICS David J. Hand
STEM CELLS Jonathan Slack
STOICISM Brad Inwood
STRUCTURAL ENGINEERING David Blockley
STUART BRITAIN John Morrill
SUBURBS Carl Abbott
THE SUN Philip Judge
SUPERCONDUCTIVITY Stephen Blundell
SUPERSTITION Stuart Vyse
SYMMETRY Ian Stewart
SYNAESTHESIA Julia Simner
SYNTHETIC BIOLOGY Jamie A. Davies
SYSTEMS BIOLOGY Eberhard O. Voit
TAXATION Stephen Smith
TEETH Peter S. Ungar
TERRORISM Charles Townshend
THEATRE Marvin Carlson
THEOLOGY David F. Ford
THINKING AND REASONING Jonathan St B. T. Evans
THOUGHT Tim Bayne
TIBETAN BUDDHISM Matthew T. Kapstein
TIDES David George Bowers and Emyr Martyn Roberts
TIME Jenann Ismael
TOCQUEVILLE Harvey C. Mansfield
LEO TOLSTOY Liza Knapp
TOPOLOGY Richard Earl
TRAGEDY Adrian Poole
TRANSLATION Matthew Reynolds
THE TREATY OF VERSAILLES Michael S. Neiberg
TRIGONOMETRY Glen Van Brummelen
THE TROJAN WAR Eric H. Cline
TRUST Katherine Hawley
THE TUDORS John Guy
TWENTIETH-CENTURY BRITAIN Kenneth O. Morgan
TYPOGRAPHY Paul Luna
THE UNITED NATIONS Jussi M. Hanhimäki
UNIVERSITIES AND COLLEGES David Palfreyman and Paul Temple
THE U.S. CIVIL WAR Louis P. Masur
THE U.S. CONGRESS Donald A. Ritchie
THE U.S. CONSTITUTION David J. Bodenhamer
THE U.S. SUPREME COURT Linda Greenhouse
UTILITARIANISM Katarzyna de Lazari-Radek and Peter Singer
UTOPIANISM Lyman Tower Sargent
VATICAN II Shaun Blanchard and Stephen Bullivant
VETERINARY SCIENCE James Yeates
THE VIKINGS Julian D. Richards
VIOLENCE Philip Dwyer
THE VIRGIN MARY Mary Joan Winn Leith
THE VIRTUES Craig A. Boyd and Kevin Timpe
VIRUSES Dorothy H. Crawford
VOLCANOES Michael J. Branney and Jan Zalasiewicz
VOLTAIRE Nicholas Cronk
WAR AND RELIGION Jolyon Mitchell and Joshua Rey
WAR AND TECHNOLOGY Alex Roland
WATER John Finney
WAVES Mike Goldsmith
WEATHER Storm Dunlop
THE WELFARE STATE David Garland
WITCHCRAFT Malcolm Gaskill
WITTGENSTEIN A. C. Grayling

WORK Stephen Fineman
WORLD MUSIC Philip Bohlman
WORLD MYTHOLOGY David Leeming
THE WORLD TRADE ORGANIZATION Amrita Narlikar
WORLD WAR II Gerhard L. Weinberg
WRITING AND SCRIPT Andrew Robinson
ZIONISM Michael Stanislawski
ÉMILE ZOLA Brian Nelson

Available soon:

IMAGINATION Jennifer Gosetti-Ferencei
DEMOCRACY Naomi Zack
BIODIVERSITY CONSERVATION David Macdonald
THE VICTORIANS Martin Hewitt

For more information visit our website

www.oup.com/vsi/

Geoff Cottrell

OBSERVATIONAL ASTRONOMY

A Very Short Introduction

OXFORD
UNIVERSITY PRESS

Great Clarendon Street, Oxford, OX2 6DP,
United Kingdom

Oxford University Press is a department of the University of Oxford.
It furthers the University's objective of excellence in research, scholarship,
and education by publishing worldwide. Oxford is a registered trade mark of
Oxford University Press in the UK and in certain other countries

Published in the United States of America by Oxford University Press
198 Madison Avenue, New York, NY 10016, United States of America

British Library Cataloguing in Publication Data
Data available

Library of Congress Control Number: 2023932759

ISBN 978-0-19-284902-1

Printed and bound by
CPI Group (UK) Ltd, Croydon, CR0 4YY

Contents

Preface

This book represents a completely revised edition of what was originally published as *Telescopes: A Very Short Introduction*, reframed to reflect the wider field of the whole observational side of modern astronomy and astrophysics, given that we no longer rely on traditional optical telescopes alone. Since *Telescopes* was published in 2016, observations are being pushed even further to the limits of what is possible. With the first ever observation of the shadow of a black hole, for example, a telescope the size of the Earth was used to see an object the size of an apple on the surface of the Moon. The development of detectors sensitive to the cosmic messages carried by neutrinos and gravitational waves now enable us to observe regions of the Universe that hitherto have been hidden from traditional telescopes. We have observed the violent collisions and mergers of neutron stars and black holes in galaxies billions of light years away via the information carried by gravitational waves. The rate of progress in this fast-moving field is extraordinary, and has shown no signs of letting up as a new generation of more powerful telescopes comes online. How is it that we modern humans, mere specks in an unimaginably vast cosmos, and who have only been around for about 200,000 years, have devised ways of seeing galaxies so remote in space and in time that they emitted their light over 13 billion years ago, when the infant Universe was only a few million years old?

List of illustrations

Chapter 1
The observable Universe

In a galaxy more than a billion light years away, two black holes, one 36, and the other 29 times the mass of the Sun, had become ensnared in each other's powerful gravitational grip, and were gyrating rapidly around each other. As the pair orbited, the spacetime around them was dynamic and distorted, producing disturbances spreading out at the speed of light across the Universe, as ripples in the fabric of spacetime—gravitational waves. As this drama played out, the black holes spiralled inwards towards their fate—the merger of two black holes into a single, bigger one. A billion years later, in 2015, the final ripples from this merger, GW150914, reached the Earth and jiggled the sensors of a new and exquisitely sensitive type of astronomical instrument, the Laser Gravitational Wave Observatory (LIGO). This was the first ever detection of a gravitational wave.

Modern observational astronomy is exciting and varied, and is capturing dramatic events in the cosmos which may well not involve traditional telescopes. Until just a few decades ago, astronomers had only photons to work with and virtually everything we know about the Universe has come from studying particles of light. Before 1945, this meant observing in the narrow visible region of the electromagnetic spectrum. Since then, great advances in technology have enabled us to probe beyond the visible spectrum, allowing us to explore the Universe at radio, infrared, ultraviolet, X-ray, and gamma ray wavelengths. These new observations, made with increasingly sophisticated technology, have shown us a plethora of exotic phenomena such as young galaxies at the edge of the visible Universe, pulsars, quasars, colliding galaxies, and exploding stars. Pulsars, or pulsating stars, are neutron stars, the relatively tiny stellar cores left behind when massive stars explode as supernovae, spinning rapidly and emitting intense beams of radiation. Quasars, or quasi-stellar radio sources, first observed at radio wavelengths, are extremely distant objects which emit vast amounts of energy and powered by matter falling into supermassive black holes. These exotic objects have been discovered using new types of telescopes that have enabled us to see farther, deeper in time, and to 'see' in different ways.

Observational astronomy is not just a story of technological development. To interpret what is observed, astronomers and astrophysicists use theory. Observations and physical theory have together enabled us to piece together how stars produce their energy, the origin of the chemical elements, how black holes form, and how supermassive black holes lurking in the hearts of galaxies can power quasars, spewing out immensely powerful jets of particles and energy thousands of light years into space. Observations have shown us that the Universe is expanding and that the expansion itself is accelerating. Sometimes we discover celestial objects that provoke us into devising ingenious ways of testing theories, and sometimes an observation is so puzzling or surprising that we are driven to revise, expand, or even replace our model of what is happening, how, and why. Either way, theory and observation are inseparably entwined, as will be evident throughout this volume.

The first maps of the stars were made in antiquity by the Babylonians and Egyptians who, 5,000 years ago, were skilled astronomers. The father of observational astronomy is a Greek, Hipparchus, who, in around 129 BC, catalogued 850 stars and invented a way to record their apparent brightness, the magnitude system, still in use today. Magnitudes range from one (the brightest) to six (the faintest visible with the naked eye), a back-to-front scale where a difference of five magnitudes indicates that one star is 100 times brighter than another.

The use of the first telescopes greatly increased our awareness of how big the visible Universe truly is. Galileo Galilei took a 3x power Dutch spyglass, multiplied the magnification 10-fold and, with it, separated the diffuse glow of our galaxy, the Milky Way, into a vast number of points of light from separate stars. As telescopes increased in size they revealed fainter objects, pushing observations ever deeper towards the boundaries of the visible Universe. Today's biggest telescopes capture 10 million times more photons than the eye, and shepherd them onto electronic retinas which are much more sensitive than those in our eyes.

Light brings us most of the knowledge we discover about the world. Nothing travels faster than light, with a speed, c, of 300,000 kilometres (km) per second. By everyday standards this speed is enormous, so that light appears to move instantaneously. However, the light reaching you from the words you are reading takes a nanosecond, one-billionth of a second, to reach your eyes. And the light from our nearest star, the Sun, takes eight minutes to reach us—we see the Sun not as it is *now*, but as it *was* eight minutes ago. The finite speed of light means that we never see the present, we always see the past. Time and space are inextricably intertwined. Astronomers have turned this into a way of measuring big distances in the Universe. The speed of light defines the astronomical distance of a *light year*, the distance light travels in one year. (One light year is almost 10 trillion km (or 10,000,000,000,000 km).) Moving out deeper in space, the nearest star to the Sun is *Proxima Centauri*. The light it emits takes four years to reach us; it is four light years away. The faint stars that Galileo saw in the Milky Way are further away still: they emitted the light that we are now seeing 10,000 years ago.

But the light that we start with is that from our star, the Sun. The Sun is a small, rather commonplace star. Its most obvious observational feature is its brightness, quantified by the solar constant, the life-maintaining flux of energy arriving at the Earth, measured as 1.37 kilowatt (kW) per square metre (m). Integrating this flux over the incident surface of the Earth, and allowing for the radiation reflected back into space, the amount of energy absorbed by the Earth is enormous and greatly exceeds all human-made power sources. The total power emitted by the Sun (the standard astronomical unit of luminosity) is 3.86×10^{26} watts, its mass (the unit of 1 solar mass) is 1.99×10^{30} kilogrammes (kg), and it is a nearly perfect sphere with a diameter of 1.39 million km.

The Sun is the most studied of all stars. It is the only star close enough to reveal surface features such as sunspots, granular surface texture, solar flares, prominences, and a hot outer atmosphere, the corona. Other stars present such small angular sizes to us that making images of the surfaces of even the nearest ones requires specialized high-resolution techniques. The light emitted by stars comes from the photosphere and is radiated as a broad thermal continuous spectrum. The surface temperatures of stars range from 3,000 kelvin (K) to 30,000K; for the Sun, it is 5,800K. (The absolute temperature scale used here is measured in kelvin starting from absolute zero, 0K, or −273 degrees Celsius (°C).) An obvious feature of the stars is that they present different colours. A star's colour relates to its temperature: hot stars radiate strongly at short wavelengths (peaking at the blue end of the spectrum), and cooler stars radiate at the red end. The Sun's spectrum peaks near the middle, in the green part. The Sun is composed of 73 per cent hydrogen and 25 per cent helium, with the remainder comprising carbon, nitrogen, and oxygen, and traces of heavier elements. (It is usual for astronomers to refer to chemical elements heavier than helium as 'metals'.) The abundances of the elements in the Sun and the gas giant planets are similar to those in the Universe as a whole.

Extensive solar observations have led to the development of what is generally known as the standard stellar model, a mathematical model that not only describes the main observational features of the Sun, but can also be applied to understanding other stars. The model includes the mechanism describing the release of energy in the Sun's hot core by thermonuclear fusion reactions, the transport of the energy to the surface by radiation and convection, and the evolution of the star. In the core of the Sun, the temperature is 15 million K, and the weight of the overlying layers press down on it squeezing the particles to densities of over 10 times that of lead. There, hydrogen nuclei (protons) collide at very high speeds. By overcoming the natural electrostatic repulsion that exists between positive nuclear charges, conditions exist that enable four hydrogen nuclei (protons) to fuse together to make helium. This process releases the nuclear binding energy that powers the Sun and all stars.

One of the most powerful tools available to observational astronomers is spectroscopy—the study of the emission and absorption of electromagnetic radiation of different wavelengths and its interaction with matter. Isaac Newton famously used a simple triangular glass prism to split sunlight into its constituent colours and demonstrated that what we have come to call 'white light' is really a mixture of light of all the colours of the rainbow. The true power of spectroscopy only began to be realized when 19th-century astronomers turned spectroscopes towards the stars. (A spectroscope is a prism combined with a small telescope designed to study a spectrum in detail.) In the Sun, hundreds of mysterious narrow dark lines (known as Fraunhofer lines) were seen superimposed on the underlying broad thermal spectrum. It was as if numerous narrow 'slices' had been removed from the spectrum. These dark lines are caused by the absorption of light at specific wavelengths, relating to the quantum energy transitions of atoms in the Sun's cooler atmosphere. The different patterns of atomic lines in a spectrum are like fingerprints revealing the presence of different chemical elements; the patterns identify a given type of atom as unambiguously as a supermarket barcode identifies its product. Spectroscopic observations therefore provided the first profoundly important evidence that the matter in the stars is the same type as matter found on Earth. Spectroscopic observations are now being made over the entire electromagnetic spectrum, and they play a central role in providing us with information about the physical state and motion of matter in distant celestial objects.

The Sun sits about halfway out from the centre of our galaxy, the Milky Way, a giant disc-like spiral galaxy, and home to 200 billion stars. Surrounding the nucleus is a central stellar bulge, embedded in a flat disc of stars and gas. If we could move out the galactic disc and look back at it, it would resemble the spiral galaxy Messier 74 (M74), shown in Figure 1. (The M74 galaxy was listed by the 18th-century French astronomer Charles Messier, a comet seeker. At that time comet hunting was a major activity and M74 was put on a watch-list of around 100 non-cometary but fuzzy objects to be avoided.) In between the stars, spiral galaxies contain a tenuous interstellar medium of gas and dust grains in the disc. From our place in the galaxy we see the Milky Way as a diffuse band of light, with dark blotches of dust clouds obscuring the light

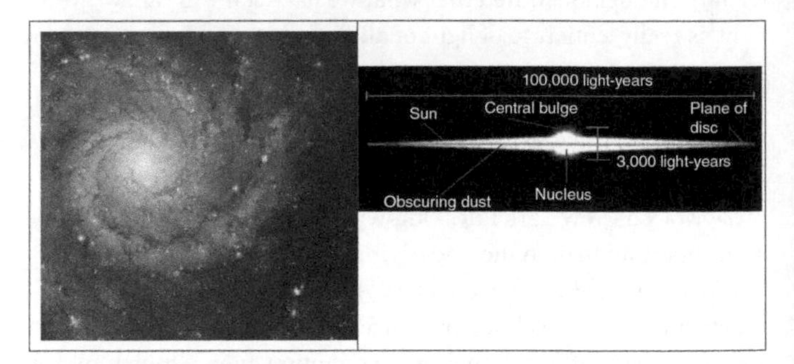

1. (Left) Hubble Space Telescope image of a face-on spiral galaxy similar to the Milky Way, Messier 74. (Right) Schematic side view of the Milky Way with approximate dimensions showing the disc, the central bulge, and the location of the Sun.

from countless distant stars. The Sun orbits around the centre of the galaxy at a speed of 230km per second, and makes one complete revolution, a galactic 'year', in 226 million Earth years.

Stars are the building blocks of galaxies, and galaxies gather into larger structures. The image of M74 gives the impression that the stars in a galaxy are densely concentrated, especially in the central bulge. But the average spacing of stars in spirals is about five light years (30 million solar diameters). This means that the chance of stars physically colliding with each other is negligible. Hence the stars in a galaxy are, in effect, collisionless. The other main class of galaxies are ellipticals, which are shaped like larger versions of the central bulges of spirals but contain little gas or dust. When we look out into space, beyond the Milky Way, we see groups and other assemblies of galaxies in the local Universe. When astronomers talk about the 'local Universe' they mean everything in our immediate vicinity for which the effects of cosmic evolution can be ignored. In practical terms this is a spherical volume roughly a billion light years across. Since the average separation of galaxies is around 10 million light years, the local Universe contains some 60 million galaxies. But when we look out further than this, we discover plenty of evidence that the Universe has undergone dramatic evolution in its history, evidence that points to its origin in the Big Bang.

The age of the Universe, namely the time that has elapsed since the Big Bang, is 13.8 billion years. It is reasonable to ask how big the observable Universe is. The oldest photons that we can observe are those of the *Cosmic Microwave Background* (CMB) radiation, the relic of an early hot and dense phase, emitted at a time before the stars and galaxies had formed. When it was emitted, the CMB radiation would have appeared to our eyes as visible light, but, owing to the expansion of the Universe, it has now been shifted in wavelength (or redshifted) by such a large factor that we now observe it in the microwave waveband. The CMB radiation has been travelling through space for nearly

13.8 billion years. We might expect that if it has been travelling for that length of time, and is only just now entering our telescopes, the visible Universe should be a sphere, centred on the Earth, with a radius of 13.8 billion light years. Correct? No, this is completely wrong.

Where this line of reasoning goes wrong is in the assumption that the Universe is *static*, namely that the distances between objects are constant. The Universe is not static. Observations have shown us that it was once smaller and hotter, and is expanding. When we measure the distance between two points on Earth, we take it for granted that we can make measurements simultaneously. But that is not the case: we are only just now seeing the CMB radiation that was emitted shortly after the Big Bang, nearly 13.8 billion years later. In the time it has taken for the radiation to reach us, space has expanded, and the place where the original light was emitted is now further away from us. The simplest calculation of the radius of the sphere around us from which the CMB was emitted suggests 41 billion light years. But even that is too small. In 1998 astronomers discovered an extra component of the Universe, dark energy, a repulsive anti-gravity energy field, filling all of space and pushing the galaxies apart. When dark energy is taken into consideration, the radius of the sphere from which the CMB light was emitted increases to 46 billion light years. So, in short: when we observe the CMB radiation with our telescopes, we are seeing it as it was when it was emitted a (relatively) short distance away; but now, because of the expansion of space, the location from which it was emitted has moved 1,000 times further away from us than it used to be. When we see this radiation, we see it as it was *then*, but not as it is *now*.

We therefore find ourselves sitting in the middle of a 46-billion-light-year-radius sphere containing all the observable matter in the Universe. There may, of course, be more matter outside the sphere, but its light has not had time to reach us yet, and so it is unobservable. The sphere of the observable Universe contains several hundred billion galaxies, equivalent to a matter content of some 10^{80} hydrogen atoms. Although it is difficult to imagine such a vast number, the Universe is so big that the *average* density of matter, spread out through the whole volume, amounts to only a few hydrogen atoms per cubic metre. By comparison, the Earth's density is almost 10^{30} times larger than that. Planets are therefore unusually dense regions of the Universe. But there are objects even denser than planets. The density of a neutron star is over 10^{13} times greater than that in a planet. These numbers begin to make sense when it is appreciated that the Universe is enormous, and consists of mostly empty space. But even then, space is not really truly empty: it contains dark energy, dark matter, the CMB radiation, subatomic neutrino particles, and highly energetic cosmic ray particles. The fabric of spacetime is also crossed by the expanding ripples of gravitational waves, emitted by the violent mergers of black holes and neutron stars.

Chapter 2
Big telescopes

Before 1945, most astronomical observations were made at optical wavelengths. By studying the visible light from the stars, nebulae, and galaxies, astronomers have amassed a vast amount of knowledge about the Universe. This chapter looks at how today's big optical telescopes have come to collect a million times more light than Galileo's instrument, how the information they yield on the structure of the Universe is being decoded, and how these powerful instruments are being used to reveal some of the most violent, distant, and ancient objects in the cosmos.

Telescopes

A telescope is effectively a light bucket, designed to gather photons and shepherd them accurately into light sensors, photographic plates, or the eyes of astronomers. One of the key attributes of a telescope is size, measured by the aperture or diameter of the main objective lens or mirror. A maxim with telescopes is: bigger is better. There are two reasons for this. First, large-aperture telescopes capture more light and so enable the faintest objects to be observed. Some of the most interesting photons come from the most distant objects in the Universe and so are sparse. A telescope focuses light onto the picture elements (pixels) of an electronic sensor to form an image or a spectrum. In general, the images of low-brightness objects are ultimately degraded by noise which can limit the scientific information being sought. There are two main noise sources: statistical noise resulting from the low numbers of photons detected, and thermal noise produced in electronic light sensors. The first of these can be reduced by capturing more photons, either by increasing the telescope aperture or the length of the exposure, or both. The second is minimized by cooling the detectors. The net effect is to increase a crucially important quantity, the *signal-to-noise ratio.*

The second reason that bigger telescopes are better is that they form sharper and more detailed images, namely they have higher angular resolution. The amount of image detail a telescope can resolve is governed by a fundamental wave property, *diffraction*, the tendency for light to bend around obstacles such as holes, sharp edges, or slits. Diffraction is commonly seen when ocean waves spread out behind a breakwater at the entrance to a harbour. Waves of any type are characterized by a *wavelength* (the distance between successive wave crests), a *frequency* (the number of oscillations that occurs each second measured in Hertz, or Hz), an *amplitude* (the height of the wave), and the *phase* of a wave (a specific location or timing of a point in a wave cycle). The related concept of *wavefronts* was first described by the 17th-century Dutch scientist Christiaan Huygens. Huygens imagined that every point on a luminous body was the source of elementary spherical *wavelets* which spread out at the speed of light. A wavefront is the surface which is tangent to all the wavelets. As a wavefront moves forward, it will itself becomes a new source of wavelets, thus describing light propagation. By throwing a pebble into a still pond, the ripples (wavefronts) spread out as concentric circles from the point of impact and, seen from a large distance away, approximate to parallel plane waves. The crests of the waves define the wavefronts which lie perpendicular to the wave direction. This is how we observe starlight; the stars are so distant that the light they emit arrives at the Earth as plane wavefronts. The smallest possible angle a telescope can resolve is called the *diffraction limit*, and is expressed in wave optics by the ratio of the wavelength of the light to the diameter of the aperture. The larger the aperture (and/or the smaller the wavelength) the higher the resolving power of the telescope.

The first telescopes were refractors, which collect and focus the light with lenses (Figure 2(a)). Refraction is the bending of light when it passes obliquely through a transparent medium such as glass or water, and is caused by the light slowing down while it is traversing the medium. A lens has the special property that when a ray strikes its curved surface, the ray is bent, and exits the lens in a new direction. The lens gathers parallel light rays from a distant

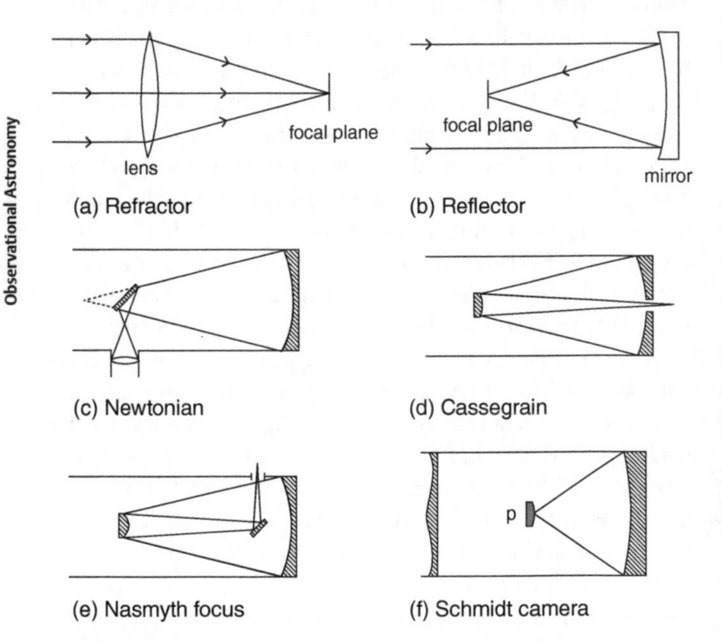

2. Different telescope designs. Top: focusing of parallel light rays to a prime focus in (a) a refractor, (b) a reflector. Below: secondary mirrors are used to produce different types of focusing configurations in reflecting telescopes ((c), (d), and (e)). The mirror in the Schmidt camera (f) is spherical, and uses a front lens to correct for spherical aberration to produce images on the curved image plane at p.

point-like object such as a star and focuses them to a bright spot on the focal plane, at a distance of one focal length away from the lens. A lens (or a curved mirror) can be thought of simply as a device that converts incoming plane wavefronts of light to spherical wavefronts, converging on a focal spot. The wavefronts arriving from the different parts of an extended object, such as the Moon, are each slightly tilted so that the light is focused on different regions of the focal plane where an (inverted) image is formed.

To convert a convex lens into a telescope that can be used visually, a second lens or eyepiece is needed, and this acts as a magnifying glass. The eyepiece converts the steeply angled rays at the focus into a near-parallel bundle of rays on which the eye can focus. Galileo's telescope of 1610 used a concave eyepiece lens placed before the main focus, and German astronomer Johannes Kepler's telescope used a convex lens placed after it. Simple refractors, however, suffer from a number of optical distortions such as spherical and chromatic aberration. By the 19th century, technical improvements largely overcame these defects, when the refractor reached its heyday as the professional telescope of choice. The largest refractor, the 1m refractor at the Yerkes Observatory in Williams Bay, Wisconsin, was built in 1897 by a visionary American astronomer, George Ellery Hale. The Yerkes telescope, however, marked the high point for refractors. The objective lens alone weighed half a tonne. Since a lens can only be supported around its edge, the glass of any larger lens would sag unacceptably, ruining the image quality.

The problematic defect of chromatic aberration which plagued early refractors was noted by many astronomers, including Newton. The edge of a lens resembles the triangular section of a prism, and behaves similarly—splitting light up into the different colours of the spectrum. Each colour has its own different focus which makes it impossible to form a single sharp white light image. With this in mind, Newton designed and built the first practical reflecting telescope in 1668 (Figure 2(c)) in which light is reflected from a concave mirror (essentially a lens turned inside out), allowing the image to be viewed with an eyepiece, and eliminating chromatic aberration. The extra flat mirror in the optical tube of the Newtonian telescope allows an observer to view the image from the side. The advantages of reflectors over refractors include being able to make mirrors with both large apertures and short focal lengths, thus effectively reducing the length and weight of telescopes. Until the end of the 19th century, however, the only material available for mirrors was a bronze alloy called speculum. Speculum is a brittle material which is difficult to cast, grind, polish, and keep from tarnishing. These drawbacks did not prevent several important reflectors being made, including those by William Herschel for his discovery of the planet Uranus (another was his largest reflector of 1789 with a 1.2m aperture) and Lord Rosse who in 1845 built the so-called 'Leviathan of Parsonstown' in Ireland. His was a giant 1.83m telescope which he used to make the first ever observation of the spiral structure in a galaxy—the 'Whirlpool' galaxy (M51). The key technical breakthrough for reflecting telescopes came in 1859 when Karl August von Steinhal and Léon Foucault discovered how to deposit silver on glass, and so opened the way to making large-diameter mirrors which could be ground, polished, and coated with silver.

The distances to the stars

Measuring distances is one of the oldest problems in astronomy. The simplest and most direct technique, trigonometric *parallax*, was well known to the ancients and is based on the surveyor's method of triangulation. The angular bearing of a distant object is measured from each end of a baseline, making a triangle with the object at the apex. This is easily seen by holding up a finger at arm's length, looking at it with first one eye closed and then the other. The finger appears to shift with respect to the background, making an angle between the two sightlines. Parallax gave the astronomers a good estimate of the distance of the Moon, now known to be a little over 380,000 km, and was also used to estimate the distance of the Sun, albeit with a little more difficulty owing to its greater distance. The Earth–Sun distance is about 150 million km, which defines the astronomical distance unit, or AU. However, because of the much greater distances to even the nearest stars, very long baselines are needed to measure stellar parallaxes. Once the Earth–Sun distance was known, it was possible to use the motion of the Earth around the Sun to observe stars at half-yearly intervals, when the Earth is on opposite sides of its orbit. The baseline is therefore 2AU. However, even the stars nearest to the solar system are so distant (the closest to us is four light years away) that their parallax angles are extremely small (less than an arcsecond) and hence difficult to measure. Measurements of the distances to the nearest stars were not made until 1838, with the high-precision telescopic measurements of German astronomer Wilhelm Friedrich Bessel.

Finding our place in the galaxy

From our place in the Milky Way, we can only see a small fraction of its 200 billion stars. We sit in what is effectively a fog bank—the view along the plane of the galaxy is obscured by interstellar dust, but when we look out and away from the plane of the disc we can see much further out to the distant Universe. In the middle of a fog bank it is tempting to believe that you are at the centre of things, which is what astronomers once thought about our place in the galaxy. However, orbiting above and below the plane of the Milky Way are hundreds of ancient globular star clusters. These are tightly bound spherical star systems, only a few light years across and densely packed with hundreds of thousands of stars. The globular clusters gather in the halo of the galaxy like bees swarming around a hive. In 1916 American astronomer Harlow Shapley observed the spatial distribution of the clusters to find out if this could indicate where the centre of the galaxy is. To make these observations, he used the world's biggest reflecting telescope, the 60-inch (1.5m) telescope sited at the Mount Wilson

Observatory near Los Angeles, California, and built by George Hale. The mirror is a paraboloid, the figure of rotation of a parabola, and the mathematical surface that reflects rays arriving parallel to the axis of the mirror (paraxial rays) to a sharp focus. Shapley saw that the globular clusters were concentrated towards the Sagittarius constellation and, from their three-dimensional (3D) distribution in space, inferred the Sun to be just over half way out (26,000 light years) from the centre of the Milky Way. For this, Shapley needed to measure the distances of the clusters. They are too far away for stellar parallax measurements, but a new method of measuring astronomical distances had just been discovered, one that depends on measuring the brightness of a special type of star whose light output varies.

The apparent brightness of a star, as measured on Earth, is very different from its luminosity, which is an intrinsic quantity defined as the total power emitted from its surface. The rate at which the emitted energy crosses a unit area on an imaginary sphere surrounding a star falls off as the square of the radius of the sphere. At twice the distance from the star there are four times as many unit areas over which the energy flux is spread, and hence the energy flux there is only a quarter as big. This is the *inverse square law* which tells us how a star's brightness drops off with distance. At one time it was thought, mistakenly, that stars all have the same intrinsic luminosities and so it might be possible to estimate a star's distance simply by measuring its apparent brightness and applying the inverse square law. In fact, the luminosities of stars vary by an enormous factor of over 10 million, from the least luminous red dwarfs to the most luminous supergiant types, such as Betelgeuse in the constellation of Orion.

In 1912, Henrietta Swan Leavitt was working in the Harvard College Observatory in Boston, USA, one of a team of human female 'computers', measuring photographic plates of variable stars in the Milky Way's smaller companion galaxies, the two Magellanic Clouds. Some stars were a pulsing type called *Cepheid* variables. The Cepheids in Leavitt's sample are all approximately at the same distance, enabling her to deduce that the brighter Cepheids have systematically longer pulsation periods, namely they obey a period-luminosity relationship. The Danish astronomer Ejnar Hertzprung soon found some nearby Cepheids which also had parallax distances and so was able to determine their luminosities and calibrate the distance scale. Measuring the pulsation period and the apparent brightness of a Cepheid variable gives its distance. This key observation extended the cosmic distance ladder out to much greater distances than the 100 light years or so that was possible by using parallax alone. What the astronomers had found was a *standard candle*, an object which, by comparing its intrinsic luminosity with its observed brightness, enables its distance (the luminosity distance) to be inferred.

At that time astronomers were debating the nature of the different types of fuzzy objects that were lumped generically together as 'nebulae'. We now know that planetary nebulae are glowing clouds of gas flung out into space by dying giant stars; reflection nebulae shine by starlight reflected by interstellar dust; emission nebulae are interstellar gas clouds whose atoms are excited and ionized by the ultraviolet light from nearby hot young stars; dark nebulae are dust clouds seen silhouetted against brighter regions; and there are also the remnants of exploding stars, or supernovae. Finally, there are *spiral nebulae*. With a telescope, some nebulae, for example the Orion Nebula (M42) in the Milky Way, have the appearance of a smooth, luminous, dusty cloud, peppered with brilliant young stars. Others, such as the great spiral in Andromeda (M31), have a smooth appearance which had at that time not been resolved into stars. Opinions on the nature of the spirals were sharply divided between two astronomers: Shapley, who believed that they are objects inside our own galaxy, and Heber Curtis, who believed them to be extragalactic star systems, or 'island Universes'. The two astronomers put forward their arguments in the so-called great debate of 1920. But with no data there was no resolution.

New data did in fact come just three years later from Hale's second large reflector, the 100-inch (2.5m) Hooker Telescope. This telescope would be the world's largest for three decades and was used to make two of the greatest discoveries in the history of astronomy. The first task was to see if any stars could be discerned in the Andromeda spiral nebula, an opportunity that was seized upon by American astronomer Edwin Hubble. The telescope did indeed reveal a powdering of extremely faint stars. While this observation suggested that Andromeda was an island Universe, it was not definitive. The crucial step came with Hubble's identification of a Cepheid variable star in Andromeda which enabled him to estimate its distance, showing conclusively that it is an external galaxy lying well outside our own Milky Way, at a distance that we now know to be around two million light years. Suddenly the observable Universe had got much bigger.

The expanding Universe

The second discovery made using the Hooker Telescope, that of the expanding Universe, is of the greatest importance to cosmology and came soon after the discovery of external galaxies. It involved the use of spectroscopy to measure the recession velocities of galaxies by observing the shift in the wavelengths of the atomic spectral lines they emit. The Austrian physicist Christian Doppler had shown in 1842 that the wavelength of any wave seen by an observer is affected by the velocity of the source of the waves, known as the *Doppler effect*. We are familiar with the Doppler effect which is responsible for the high-to-low change in pitch of an ambulance siren as it zooms past us. When an ambulance is approaching, the peak of each sound wave is emitted from a position closer to the observer than the previous one, and so takes less time to arrive. The motion bunches the wave peaks closer together, which increases the pitch. As the ambulance speeds away, the peaks are stretched out, and the pitch drops. In 1868, the British astronomer William Huggins had been the first to put this into practice and measure the radial velocity of a star,

Sirius, by comparing the wavelength shift in its spectral lines with laboratory measurements. The light from a star that is receding from us at a velocity v (much less than the speed of light, c) shows spectral lines that are shifted towards the long-wavelength red end of the spectrum, by an amount given by the *redshift*, $z = v/c$. If the star is approaching, the light is *blueshifted* towards the blue end of the spectrum.

By 1917, the American astronomer Vesto Slipher had extended this technique to measure the radial velocities of entire galaxies whose spectrum consists of the merged light of billions of stars. Such measurements are challenging because galaxies are low-brightness objects. Moreover, when the diffuse galactic light is dispersed and recorded by a spectrograph (a camera combined with a spectrometer) the amount of signal in each wavelength interval is further diluted. To achieve a sufficiently high signal-to-noise ratio for these measurements, Slipher used a large-aperture telescope and long-exposure photographs of the spectra. This showed that distant galaxies are mostly redshifted, namely that most of them are receding from us. Edwin Hubble and Milton Humason later used the Hooker Telescope to observe a sample of nearby galaxies for which they had both redshift and distance information. This showed that the galaxy redshifts increase systematically with distance. Wherever you look out into the Universe, galaxies appear to be receding away from us with speeds, v, proportional to their distances, D. This relationship indicates an expanding Universe described by the *Hubble law*, $v = H_0 D$. The constant of proportionality is the Hubble constant, H_0, which is a measure of local rate of expansion. (The law is now called the Hubble–Lemaître law, acknowledging the theoretical predictions of an expanding Universe by the Belgian priest and cosmologist Georges Lemaître.) At first sight, the Hubble law suggests that we occupy a privileged position in the Universe. But it turns out that an observer on any other galaxy would observe the same phenomenon. When the discovery was announced in 1929, it hit the astronomical community like a bombshell. At one stroke, it overturned the notion of a static Universe, and implied that the Universe is evolving, and that it had a definite beginning. Think of a movie showing the galaxies flying apart from each other, and then run the movie backwards in time. At a certain time in the past all the galaxies merge together in a state of infinite density in the *Big Bang* at the beginning of the Universe.

George Hale's last big telescope, the giant 200-inch Hale Telescope, sits in a magnificent white dome the size of the Pantheon in Rome at the Palomar Observatory in San Diego county where it saw first light in 1948. The telescope's enormous 5.1m mirror proved to be challenging to fabricate. The 13-tonne, low-expansion Pyrex glass blank took 11 years to grind and polish, and was figured as a paraboloid to within 1,000th of the width of a human hair. The heavy monolithic mirror relies partly on its own stiffness, and partly on the back support of a massive steel cell to maintain its shape under gravity when tilted over a wide range of angles. At the telescope's prime focus sits an instrument cage which has carried observers moving with the telescope as it tracks the sky. There is a hole in the primary mirror for a Cassegrain focus (Figure 2(d)) produced using a hyperboloidal secondary mirror. The Hale Telescope was the world's largest reflector for 45 years. It has played a crucial role in elucidating important astrophysical phenomena and it has been used to measure the Hubble constant more precisely than before; to study the isotropy and linearity of the expansion of the Universe; and to contribute data for the theory of stellar nucleosynthesis, the building of the chemical elements by fusion reactions. The telescope has also identified the different stellar populations in galaxies, which has greatly advanced our understanding of galaxy evolution. Although it is no longer the world's largest telescope (that honour passed to the two Keck telescopes in 1992), the Hale has now been equipped with an adaptive optics system to increase the angular resolution and a high-resolution spectrometer, and is now being used to probe the atmospheres of exoplanets.

New big telescopes

The apertures of the biggest optical telescopes have doubled every 40 years. To maintain that rate of progress, new concepts and technologies have had to be incorporated into the latest generation of 8–10m telescopes. Innovations have been made in the development of telescope mounts, new ways of making big mirrors and efficient light sensors, and sophisticated techniques of removing the blurring of telescope images caused by turbulence in the Earth's atmosphere.

In the simplest type of *transit telescope* mount (Figure 3(a)), a telescope is pivoted about a horizontal east–west axis, tilting the telescope up at different altitude angles above the horizon along the great circle line of the meridian, the line joining the celestial poles and passing through the observer's zenith. Observations are made of strips of sky drifting through the field of view with the Earth's rotation, and the positions of objects are recorded by noting their crossing times and altitude angles.

With the 19th-century invention of photography and its almost immediate use in astronomy, celestial objects could for the first time be recorded using long exposures. To avoid image blurring, it soon became essential for telescopes to track objects accurately across the sky. The traditional *altitude-azimuth* (alt-az) mount (Figure 3(b)) has the same geometry as a gun turret, and can point a telescope to anywhere in the sky. As in the transit mount, the telescope pivots in elevation but is also mounted on a rotatable horizontal platform allowing it to be pointed to any azimuth angle (or compass bearing) and so to anywhere in the sky. To track a sky target requires simultaneous rotations about the altitude and azimuth axes, as well as a rotation of the imaging sensor, all with different rates. Before computer control, it was mechanically impossible to perform this complex set of rotations to the required precision, and a different system was developed—the *equatorial*

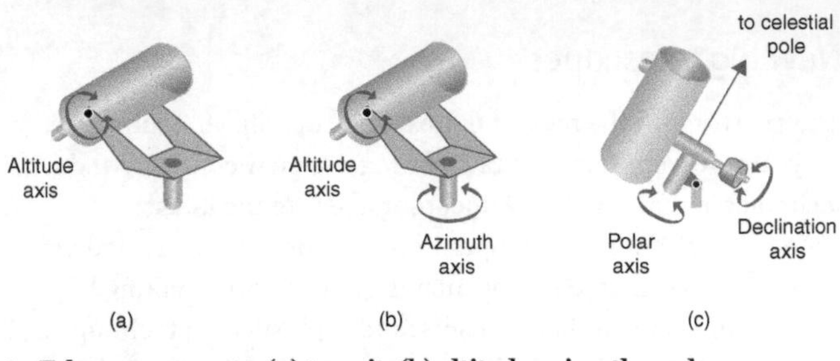

3. Telescope mounts: (a) transit; (b) altitude-azimuth; and (c) equatorial.

mount (Figure 3(c)). If the base of an alt-az mount is tilted up so as to align the azimuth axis with the celestial pole, stars can be tracked by steadily rotating the telescope about just one axis (the polar axis). This simplification was achieved at the expense of mechanical complexity which, for large observatory telescopes, was significant. All of Hale's telescopes were mounted equatorially but, with the advent of computer control, modern observatories (including moveable radio telescope dishes) have now adopted the alt-az system.

Big mirrors

Large modern telescopes have primary mirrors with geometries different to Hale's paraboloids. A paraboloid focuses incoming paraxial rays to a sharp spot, but the peripheral rays become smeared out into a distorted blob which grows a tail, or coma, like a comet. This so-called *comatic aberration* worsens the further an object is from the axis, and ultimately it limits the usable field of view. Many modern telescopes use the wider field optics of the Cassegrain *Ritchey-Chrétien* design (Figure 2(d)), which is based on hyperboloidal primary and secondary mirrors. Modern telescopes now also have smaller focal ratios. The focal ratio (or f-number) is the ratio of the focal length of the mirror to its diameter and is a measure of the 'speed' of the optics. The name derives from photography: the lower the f-number, the faster the optics, and the shorter the exposure time needed to form an image. The trend over the last 50 years has been for f-numbers of telescopes to get smaller, and values below f/2 are now common.

The Hale Telescope has the largest thick solid-glass mirror that it is possible to make and, since the 1970s, designs have concentrated on thinner and lighter mirrors. Thin mirrors are floppy which means that they need active mechanical support to stay in shape. One variety, *meniscus mirrors*, are used in European Southern Observatory (ESO) Very Large Telescopes (VLTs), shown in Figure 4. Each of the VLT's 8.2m, 17.5cm thick mirrors is supported by an 'active optics' system using 150 electro-mechanical actuators supporting its back which correct the mirror figure in real time by computer-controlled feedback.

An alternative design is the *segmented mirror*, in which one large mirror is constructed from a number of smaller hexagonal segments, each part of the parent mirror shape. This technique was pioneered in the two 10m Keck telescopes (Figure 5), which are sited 80m apart on top of the extinct volcano Mauna Kea in Hawaii, 4,145m above the Pacific Ocean. Each mirror segment is 1.8m across, 7.5cm thick, and computer-aligned. The Keck telescopes can

4. The European Southern Observatory Paranal Observatory, 2,635m above sea level, in the Atacama Desert in Chile. The four large buildings house the four 8.2m Very Large Telescopes.

5. A Keck telescope primary mirror, showing the hexagonal segments.

be operated individually or as an interferometer pair to increase their angular resolution. The largest segmented-mirror telescope currently operating is the *Gran Telescopio Canarias* (GTC), located in La Palma (altitude 2,267m), with an aperture of 10.4m. The first space telescope to use a segmented mirror is the 18-segment, 6.5m mirror of the *James Webb Space Telescope* (JWST).

Effects of the atmosphere

Most of the Earth's atmosphere lies within 16km from the surface, and is composed mainly of nitrogen and oxygen. It contains traces of water vapour, carbon dioxide, and other gases. Further out, the air gets thinner until it merges with outer space. In the *ionosphere*, a layer 75–1,000km high, atoms are ionized by solar radiation and high-energy cosmic ray particles from distant parts of the

Universe. While the atmosphere is an essential part of the biosphere, protecting life from harmful cosmic radiation, it also filters the wavelengths of the electromagnetic spectrum that reach the ground from space. The incident radiation spans a huge range of wavelengths in gamma rays, X-rays, ultraviolet (UV), visible light, infrared (IR), millimetre (mm) radiation, and radio. The transmission curve of radiation through the atmosphere is shown in Figure 6. Incoming high-energy gamma ray, X-ray, and UV photons have enough energy to tear electrons from atoms in the atmosphere, ionizing the gas and rendering the atmosphere opaque at these wavelengths. Below the violet end of the visible spectrum, the sharp increase in opacity is caused by the strong absorption of near-UV photons by atmospheric ozone (O_3).

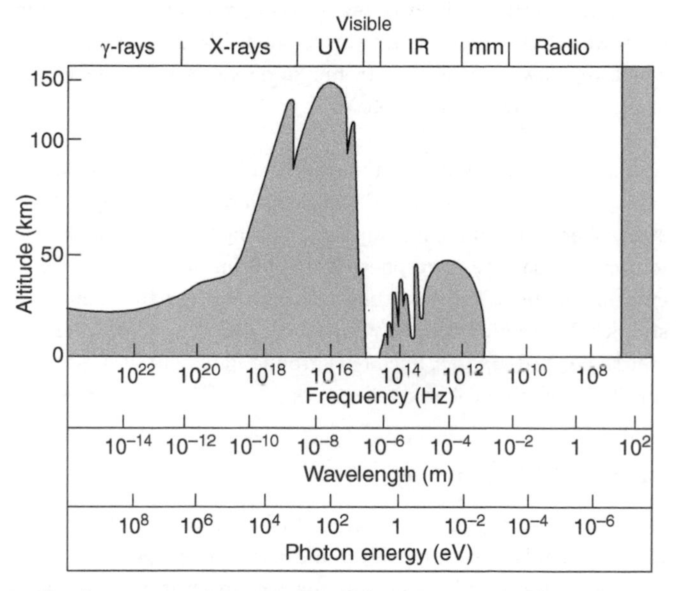

6. The electromagnetic spectrum and the transparency of the atmosphere. The shaded region is the altitude at which the atmosphere is transparent to cosmic radiation of different wavelengths.

At the longer wavelengths of infrared, water and carbon dioxide molecules absorb most of the radiation. These molecules absorb and emit infrared photons in a multitude of narrow vibrational lines, and it is these that produce the complex shape of the opacity curve. For wavelengths between about 1cm and 10m, a second transparent window opens up in the atmosphere which is used for radio observations. At even longer wavelengths, the incoming radiation is reflected back out into space. At such low wave frequencies, the ionospheric electrons are mobile enough to screen out the oscillating wave fields causing the ionosphere to act like a mirror. Earthbound short-wave radio operators exploit this mirroring effect when they communicate with stations over the horizon, by transmitting radio signals that bounce off the inside surface of the ionosphere. In fact, the Earth's surface and the ionosphere behave rather like the walls of a waveguide (familiar in microwave technology). These surfaces define a cavity within which very low frequency electromagnetic waves are trapped and can propagate around the globe.

Although the atmosphere is transparent in the visible one-octave (400–800 nanometres, or nm) wavelength window, by night, the atmosphere is never completely dark. During the day, sunlight ionizes atoms in the atmosphere which recombine at night, emitting photons. The atmosphere also generates light when struck by energetic cosmic ray particles, producing a low-level emission called *airglow*. Such background light sources limit the sensitivity of ground-based telescopes.

Atmospheric blurring

At optical wavelengths, atmospheric turbulence defines the effective angular resolution of ground-based telescopes. Turbulence is driven by winds mixing cool and warm air streams generating high- and low-density air pockets which cause the bumpiness experienced by air travellers. Air refracts light by a small amount which, over the 100km-long path of photons arriving from outer space, makes the stars twinkle. The flat wavefronts arriving from distant objects are distorted in their passage through turbulent cells in the atmosphere. Observing the stars through the dynamically fluctuating atmosphere is akin to trying to see a coin lying on the bed of a stream of rippling water.

The optical quality of the atmosphere is measured by a quantity called *seeing*, the degree to which the atmosphere smears out the point-like image of a star into a seeing disc. Seeing discs are typically one or two arcseconds across and, at the best observing sites, which are generally at high altitude, can be as small as 0.5 arcseconds. However, even this size is 20 times coarser than the ideal diffraction limit of a 4m telescope. It turns out that the resolving power of a telescope with an aperture any larger than about 20cm is fixed by atmospheric seeing and not by the optical aperture. Despite this, bigger telescopes collect more photons enabling fainter objects to be observed.

Modern observatory-class telescopes counter the atmospheric blurring with a wavefront-correction method called *adaptive optics* (AO). AO removes the atmospheric distortion by restoring and flattening out wavefronts. The two key elements in an AO system are an optical wavefront detector (which measures the phase distortions of the incoming light), and a dynamically deformable mirror to apply corrections in real time. The wavefront detector is essentially an array of tiny telescopes, each of which measures the shift in position of a bright reference star from which the pattern of atmospheric distortion is computed. The deformable mirror is flexed in real time by piezoelectric actuators attached to its back and a computer feedback system restores the plane parallel wavefronts. An AO system relies on the presence of a bright point-source reference star lying in the field of view. If no suitable star happens to be there, an 'artificial star' is created by firing a laser beam from the telescope, tuned to excite sodium atoms at the top of the atmosphere.

Recording images

The development of photography in the mid-19th century revolutionized observational astronomy. Before photography, astronomers had only their eyes to observe with, and what was seen through the telescope could only be recorded in the form of hand-drawn sketches. Apart from eliminating human bias and error, photography has the added benefit of overcoming a natural limitation of the human eye. Similar to a movie camera, the eye makes a sequence of short exposures about 20 times a second. Such a refresh rate is good for detecting motion, but it limits the number of photons captured, and hence sensitivity. A long photographic exposure not only reveals much fainter objects, but it also accurately captures the positions and magnitudes of many of them at the same time, establishing an objective record of the sky.

A dramatic improvement in the efficiencies of telescopes has come through the development of sensitive electronic light sensors which now replace the photographic plates that were in common use until the early 1980s. Electronic detectors are sensitive to a wider range of wavelengths than either the eye or photographic emulsion. Light is detected by the photoelectric effect in materials, the transfer of energy from photons to release photoelectrons, producing a small electrical current. The efficiency of a light sensor is defined by the *Quantum Efficiency* (QE), namely the fraction of incoming photons converted into useful signals. The QE of the eye and the light-sensitive grains in a photographic emulsion is only a few percent. In 1969, Willard Boyle and George Smith of Bell Laboratories in New Jersey invented a new type of light sensor, the *Charge Coupled Device* (CCD). CCD sensors have QEs approaching 100 per cent and have revolutionized astronomical imaging. Simply by replacing the photographic plate with a CCD sensor the sensitivity of a telescope can be boosted about 40 times. In terms of light-capture efficiency, it is as if the telescope's aperture had been increased six-fold.

A CCD detector is a silicon chip that counts the photons that fall on a chequerboard array of micron-sized photosensitive cells (photosites). When a visible light photon strikes a photosite, a photoelectron is released and stored. As more photons rain down on the chip, the photosites fill up with electrons, collecting like raindrops in buckets. An image from a telescope is represented as a 2D pattern of electrical charge in the CCD. After an exposure, the accumulated charges are read out from the pixels and stored digitally. The CCD is also a *linear* detector. The quantity of charge stored in a pixel is proportional to the number of photons that arrived during an exposure, a feature that is important in photometry, measuring the brightness of objects. CCD cameras are generally cooled to low temperatures to reduce thermal noise.

Many interesting objects in the Universe emit more infrared than visible light. These are cold objects such as brown dwarfs, planets, dust clouds, and galaxies near the edge of the Universe for which the visible light emitted has been redshifted to longer wavelengths. Infrared is less strongly absorbed by interstellar dust than visible light; infrared allows the interiors of dust-shrouded objects to be observed and enables us to see deep into the dusty disc of the Milky Way. The optical window in the atmosphere extends the visible part of the spectrum to longer near-infrared wavelengths (0.8 to 2.5 microns) and large observatory telescopes operate at both visible and near-infrared wavelengths. The pure silicon sensors in CCD cameras are sensitive to light with photon energies larger than about 1.1 electron volt (eV) (corresponding to a longest wavelength of 1.1 microns). However, at 2-microns wavelength, in the near-infrared, the photons only have 0.6eV which is not enough to release photoelectrons from silicon. (The eV is a convenient energy unit for the atomic scale where 1eV is the energy needed to move an electron through a potential difference of 1 volt.) To observe at longer wavelengths, alternative semiconductor materials have been developed, such as alloys of mercury, cadmium, and tellurium (HgCdTe) which are used in infrared sensor arrays such as the Near Infrared Camera (NIRCam) on the JWST.

The supermassive black hole in the centre of the Milky Way

The interstellar dust clouds in the disc of the Milky Way contain micron-sized grains of graphite and silicates. Since these grains are larger than the wavelengths of visible light they both absorb and scatter light. When light interacts with a dust grain, the oscillating electric fields force its electrons to vibrate in sympathy, turning the grains into quivering electric dipoles (spatially separated positive and negative charges) that reradiate the light energy, scattering it in all directions like miniature broadcasting stations. Importantly, the scattering efficiency is much higher at visible wavelengths than it is in the near-infrared. Observations made in the infrared therefore allow observers to see through all the obscuring dust, all the way to the centre of the Milky Way, and were therefore key to making the observations that confirmed the presence there of a supermassive compact object. The centre of the galaxy contains a compact radio source called Sagittarius A* (Sgr A*) and it was this region that had been suspected of harbouring a supermassive black hole.

A black hole is a highly compact massive object, with all the mass concentrated at a single point, a singularity, where the density is unimaginably high and the laws of classical physics break down. The singularity is unobservable because it is enclosed by an *event horizon*, a surface that cuts off a region of spacetime from the outside world. Near the event horizon the gravitational field is so strong that Einstein's General Theory of Relativity is needed to describe the weird warping of space and time that occurs. In a non-spinning black hole, the radius of the event horizon is known as the *Schwarzschild radius*, named after the German physicist Karl Schwarzschild. The Schwarzschild radius can be defined for any massive body and has the property that if all the mass of the body were to shrink down into a sphere of that radius, the escape velocity from the surface would reach the speed of light and so nothing, including light, can escape from it. If you were watching a clock falling into a black hole, the clock would appear to tick slower and slower as it approached the event horizon. The image of the clock would also get dimmer and redder and, as it reached the horizon, the time indicated on its dial would be frozen as it faded from view. This effect is the *gravitational redshift* which occurs because the photons emitted by the object lose energy as they climb out of the deep gravitational potential well. Astronomers estimate there to be around 100 million black holes in the Milky Way; these result from the collapse of massive stars, with masses as high as 150 solar masses. However, the supermassive black hole at the centre of the Milky Way is much more massive than any of these stellar corpses.

If a black hole is truly 'black' and emits no light, it is reasonable to ask how it can be observed at all. The method that was used to identify the Sgr A* supermassive black hole was to make observations of the orbits and speeds of individual stars orbiting close to it. Owing to the black hole's extreme mass and compactness, stars close to it attain exceedingly high velocities. Studying the orbits of these stars therefore informs us on the mass of the invisible central object. The galactic centre is, however, also crowded with numerous stars and star clusters. Therefore, to rule out other confounding effects, it was necessary to measure individual stellar orbits as close to the centre as possible. The key idea is that the velocities of stars orbiting in the immensely strong gravity of the black hole should be extremely high (much larger than if they were orbiting around star clusters for example). The observations needed a very high angular resolution, and the orbits of stars were followed for more than three decades using the Keck and the VLT telescopes equipped with AO at a wavelength of 2.2 microns (Figure 7). The results clearly indicated the presence of a supermassive black hole with a mass of 4 million solar masses. One star in particular, S2, has a highly eccentric orbit and a short

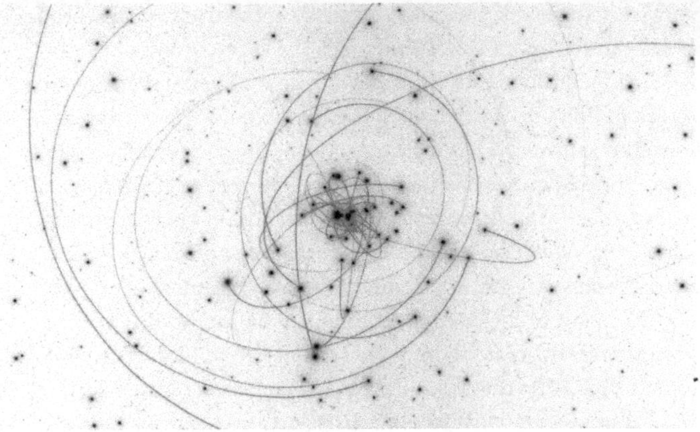

7. Stellar orbits near the supermassive black hole Sagittarius A* at the centre of the Milky Way.

orbital period of 16 years. S2 came within 120 AU of Sgr A* in 2018, where it made its closest approach to the black hole at a speed of 7,600km per second (3 per cent of the speed of light), making it the fastest known ballistic orbit so far recorded. S2 provides an excellent laboratory to test Einstein's General Theory of Relativity. To that end, spectroscopic observations have detected the gravitational redshift of S2's spectral lines, an effect that had never before been seen in a star near a black hole.

Sky surveys

Sky surveys are the foundation of observational astronomy. The aim is to collect data from well-defined areas of the sky, wavelengths, and limiting sensitivities with no specific prior astronomical targets. Having captured, digitized, and entered the data into a database, a large number of studies can be undertaken, such as cross-correlation with surveys made at other wavelengths or observation times. Astronomical surveys offer three types of information: investigation of the statistics of different types of objects in the Universe, a way of discovering new types of object, and the means to select objects for further observations with big telescopes. The objects in the database can also be presented to millions of citizen scientists for classification in online projects such as *Galaxy Zoo*.

The sky occupies a total of 41,254 square degrees, equivalent to 206,270 full Moons. This is a huge area for a telescope to cover. The first big telescopes of the 20th century typically had fields of view of around 1°, which meant they were good for seeing detail in small regions of the sky but made poor survey instruments since it would have taken an enormous time for them to complete a whole-sky survey. The need for wide-field auxiliary survey telescopes with fast optics was recognized early on and was met by a radically different type of reflecting telescope.

The *Schmidt camera* (Figure 2(f)) was invented in 1930 by an Estonian optician, Bernhard Schmidt, who abandoned the then-usual paraboloidal mirror for a spheroidal one which is much easier to make. All spherical mirrors (and lenses) suffer from the optical defect of spherical aberration, where the outer marginal rays are reflected (or refracted) too steeply for the image to be in focus. To correct for this, Schmidt added a weakly refracting corrector plate to the front of the telescope. A telescope that combines reflecting and refracting elements is called a catadioptric telescope. The large mirror-diameter-to-focal-length ratio (small f/number) of a Schmidt camera offers fast optics and a wide field of view, and is therefore ideal for making surveys of the sky. The prototype 18-inch (0.46m) Schmidt camera at the Palomar observatory was used very effectively by Swiss astronomer Fritz Zwicky from 1936 to discover many supernovae. This was followed in 1950 by the famous Samuel Oschin 48-inch (1.2m) Schmidt camera which has a 47 square degree field of view (equivalent to 200 full Moons), and could complete a survey of the northern sky in four years. The image plane in this telescope is curved, and so specially curved glass photographic plates had to be used. The survey plates it produced for the National

Geographic Society–Palomar Sky Survey continue to provide a valuable resource, and the survey was digitized in 1994.

Galaxies are building blocks of the Universe, and galaxy surveys provide information on one of the foundations of physical cosmology—the study of the large-scale structure of the Universe. The simplest type of observation is the study of the positions of galaxies on the sky, plotting them as millions of points on the celestial sphere. The results show that galaxies are not distributed randomly across the sky, but are clumped on the different scales of the *cosmic web*. If, in addition, the redshift of each galaxy is also measured (making it a *redshift survey*), then the Hubble law can be used to estimate the galaxy distances, so revealing their 3D distribution in space. An example is the *Sloan Digital Sky Survey* (SDSS) which comes from a robotic 2.5m telescope located at Apache Point, in New Mexico. Figure 8 shows an SDSS redshift survey plotting galaxies over a quarter of the sky. In practical terms, measuring spectroscopic redshifts for each of a million galaxies would be extremely time consuming. Astronomers have therefore devised a much quicker method—estimating them photometrically. Photometric redshifts are based on approximating the visible light spectrum of a galaxy by observing it with a few (around five) relatively broad wavelength optical filters. The intensities measured in each of these yield data from which the redshift can then be extracted by comparison with reference templates.

The cosmic web shows the hierarchical distribution of galaxies on different scales. It shows giant superclusters of galaxies, clusters containing hundreds to thousands of galaxies, all the way down to smaller groups of a few tens of galaxies, such as the Local Group, to which our Milky Way galaxy belongs. The clusters are interlinked by filaments and dense walls of galaxies and, between the clumps, there are vast empty voids, up to 300 million light years across.

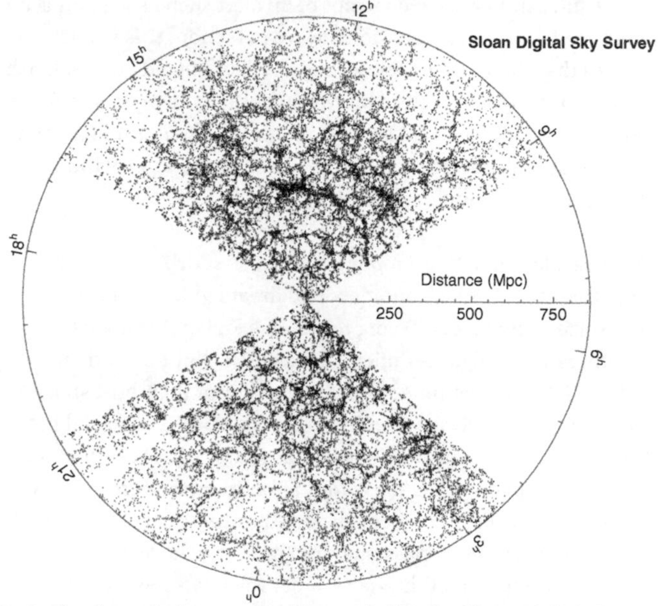

8. A slice through the cosmic web revealed by the Sloan Digital Sky Survey. Earth is at the centre and each point represents a galaxy, with its distance from the centre of the circle plotted out to 2.8 billion light years in the local Universe. The region between the two wedges was not mapped because of dust obscuration in the Milky Way.

Robotic telescopes and the discovery of extrasolar planets

Smaller telescopes have many important roles. Groups of telescopes can be spread over different sites and operated robotically to form effectively a single instrument. These are coordinated in global networks which can respond rapidly to astronomical transients. For example, the *Las Cumbres Observatory Global Telescope Network* consists of 25 autonomous telescopes with apertures between 0.4m and 2m, distributed over sites in Australia, South Africa, Tenerife, Texas and Hawaii in the

USA, Chile, and Israel. On receipt of an alert such as a gamma-ray burst, neutrino, or a gravitational wave detection, this system ensures that there will be telescopes on the night side of the Earth available for follow-up observations in both hemispheres. A fully robotic telescope, such as the 2m *Liverpool Telescope* on La Palma, can, for example, be pointed at a new transient source within minutes of the receipt of an alert.

An extrasolar planet, or *exoplanet*, is a planet orbiting a star other than the Sun. Astronomers had dreamed for centuries of discovering exoplanets. Trying to see an exoplanet using its directly reflected light is difficult because the light from their surfaces is billions of times fainter than that of their host stars. It is akin to trying to see a firefly buzzing around a powerful but distant searchlight pointing straight at you. Indirect methods of detecting exoplanets like the *radial velocity method* offer a much easier route. When a planet orbits its star, they both orbit around the common centre of mass, making the star wobble a little. The periodic modulation of the star's radial velocity shows up as a regular shifting of its spectral lines, blueshifted when the star is approaching, and redshifted when receding. The first exoplanets ever detected were observed at radio wavelengths, by Polish astronomer Alex Wolszczan and Canadian astronomer Dale Frail in 1992. They discovered exoplanets orbiting around a collapsed star—a pulsar (PSR B1257+12). A pulsar is a rapidly rotating type of neutron star emitting regular clock-like blips of radiation. The arrival times of the pulses from PSR B1257+12 were observed to vary systematically indicating the presence of three exoplanets.

The first detection of an exoplanet orbiting a normal star 51 Pegasi (51 Peg b; Figure 9) was made three years later by Didier Queloz and Michel Mayor using the 1.9m telescope at the Observatoire de Haute Provence in France. They used a high-resolution spectrometer to measure radial velocities with a precision of a few metres per second. The orbit of 51 Peg b revealed two remarkable facts: the exoplanet has a large Jupiter-like mass, and its orbital

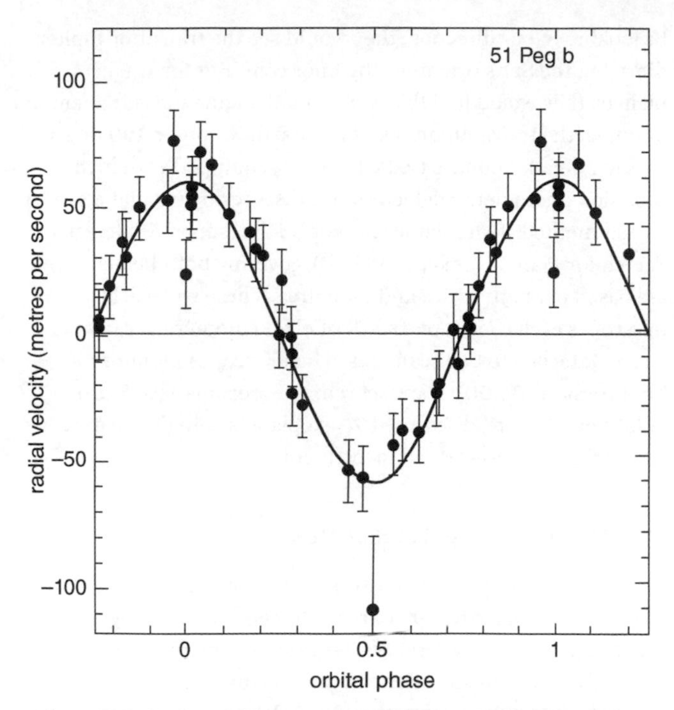

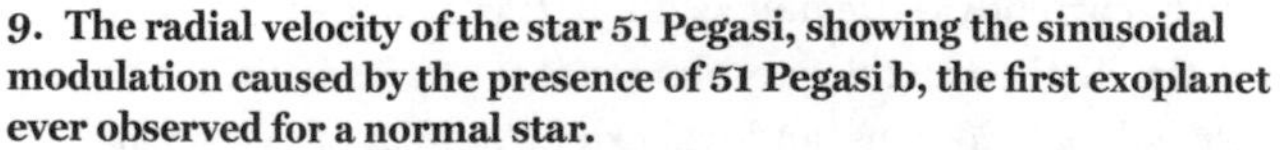

9. The radial velocity of the star 51 Pegasi, showing the sinusoidal modulation caused by the presence of 51 Pegasi b, the first exoplanet ever observed for a normal star.

period is only 4.2 days, indicating that it orbits closer to its star than Mercury does to the Sun. This was the first hot Jupiter to be observed, and a spate of observations and exoplanet discoveries soon followed. Over 5,000 or so exoplanetary systems have now been discovered, and a significant fraction of these are hot Jupiter types.

A different method of detecting exoplanets is the *transit method*, which relies on the tiny dimming of the light from a star when a planet passes in front of it. It is relatively easy to detect large planets this way. If, for example, an alien astronomer happened to be looking in our direction, they would see the transit of Jupiter dimming the Sun's brightness by 1 per cent. But for smaller planets, they would find this a more challenging measurement; for example, the transit dimming of the Earth would be 100 times smaller. Small ground-based telescopes, equipped with high-precision photometric detectors, are discovering exoplanets by the transit method. Two robotic telescopes, the (super) Wide-Angle Search for Planets (or super-WASP), covering both hemispheres, are based in South Africa and La Palma. These wide-field telescopes each consist of a stack of eight commercial camera lenses, attached to CCD cameras. The telescopes monitor the brightness of 100,000 stars, searching for exoplanets. Super-WASP has, to date, discovered 75 exoplanets, a high scientific return for a relatively small investment.

Two future optical telescopes

The largest next-generation infrared and optical telescope is the ESO's European Extremely Large Telescope (E-ELT; Figure 10). The E-ELT will have a 39m (f/1) segmented mirror, and is currently under construction at Paranal Observatory in Chile with first light currently estimated as 2028. It is designed to achieve a diffraction limited angular resolution of 0.01 arcsecond at a wavelength of 2 microns, and will extend studies of exoplanets and their atmospheres, searching for the spectral signatures of biomarkers including ozone, carbon dioxide, and water vapour, as well as looking for the first galaxies, and studying supermassive black holes and dark energy.

The *Vera C. Rubin Observatory* (previously the Large Synoptic Survey Telescope) is currently being built in Chile with first light expected by 2024. A novel feature about this powerful 8.4m robotic survey telescope is a three-mirror design which will produce a wide field of view (3.5°; equivalent to 50 full Moons). The camera consists of a mosaic of 189 16-megapixel CCDs, making it the largest digital camera ever built, about the size of a car.

10. Artist's impression of the European Extremely Large Telescope. Note the relative size of cars.

The observatory will survey the same half of the southern sky (10,000 square degrees) every three days. This high cadence will enable it to make 'movies' of the sky and discover many transient and moving objects, such as supernovae, near-Earth objects, gamma-ray burst counterparts, and variable stars. The databases will contain about 20 billion galaxies, and a similar number of stars. It will also measure the gravitational lensing of dark matter and create maps showing the evolution of dark matter.

This chapter has introduced some of the exciting advances in astronomy and astrophysics made by ground-based optical and near-infrared telescopes and the new technologies that have made the observations possible. However, these wavebands occupy only a small segment of the electromagnetic spectrum and, next, we will keep our feet on the ground and our eyes on the stars, but look at them with wavelengths a million times longer.

Chapter 3
The radio Universe

Radio emission from the cosmos was first observed accidentally in 1933 by an engineer working for Bell Telephone Laboratories in New Jersey. Karl Jansky had been trying to track down a radio hiss that had been interfering with a transatlantic radiotelephone service and, to locate the interference, had constructed a rotatable directional antenna tuned to a wavelength of 15m. The radio noise appeared to be coming from one direction which, over the course of a year, changed steadily by 1° per day (moving with the stars at the sidereal rate). This indicated it had a celestial origin, which Jansky identified as coming from the Milky Way, an observation that makes him the first radio astronomer. Jansky is twice honoured by having the basic astronomical unit of radio flux density named after him, as well as one of the world's largest radio telescopes, the Karl G. Jansky Very Large Array (JVLA) telescope in New Mexico.

Jansky's discovery was followed up a few years later by a radio enthusiast, Grote Reber. Using a 9m home-made parabolic dish in his Illinois back garden, Reber mapped the radio noise from the Milky Way at a wavelength of 2m. His dish was mounted as a transit telescope, which he used to record the brightness of the radio sky in the form of drift scans made at different altitudes. When the data were plotted as a contour map, it clearly showed the radio emission coming from the plane of the Milky Way, with the peak of intensity centred on the constellation of Sagittarius.

The question was: what is the source of this galactic radio emission? All bodies radiate electromagnetic radiation, simply because they are hot. Stars are a good example of this. Thermal radiation has a broad spectrum of wavelengths, the Planck spectrum, after the German physicist Max Planck who, in 1900, derived the formula for thermal radiation by postulating that light energy is quantized. Electromagnetic radiation is produced whenever electrical charges are accelerated, and those in a hot body are constantly in a state of random agitation; the radiation from such a mêlée of jiggling charges is emitted with many wavelengths. The hotter the body, the more violent the agitated motion, and the wider the range of frequencies that are emitted. A body emitting a Planck spectrum peaks at a characteristic frequency which depends solely on its temperature. Think of a chunk of iron heated up to forging temperatures in a blacksmith's workshop. At first it glows dull red, then bright orange-yellow, and finally a brilliant blue-white colour. (Even when cooled to room temperature, the iron continues to radiate, but its emission peak now shifts to long infrared wavelengths to which our eyes are not sensitive.) It turns out that the colour of a hot body can be used as a proxy for its temperature (hence the term 'colour temperature'). This is why it is possible to estimate the temperature of stars from their colours.

However, the radio noise emitted by the Milky Way does not have the same form of thermal spectrum emitted by stars, nor does it have a discrete line spectrum, such as that emitted by excited atoms. Instead, the radio noise has a continuous spectrum, and Reber showed that it was much stronger at low frequencies than at high frequencies. The Milky Way's radio emission was also found to be *polarized*, and this gave an important clue to its origin. In an electromagnetic wave, the electric and magnetic fields oscillate in directions perpendicular to the direction of propagation. If the radiation source is a hot body, the field directions are uncorrelated and the light is unpolarized. For the cosmic radio waves to have been polarized, they must have been produced by a non-thermal process. The mechanism was *synchrotron radiation*, a type of radiation emitted by fast electrons deflected into helical corkscrew paths by magnetic fields. This has the effect of aligning the oscillating wave fields in one particular direction. Synchrotron radiation is produced in laboratory particle accelerators such as cyclotrons and synchrotrons, and the mechanism is now recognized to operate in a wide variety of cosmic sources. In an *active galactic nucleus* (AGN), for example, electrons are accelerated to relativistic energies, causing them to gyrate in magnetic fields and emit polarized radiation. In the Milky Way, electrons are also accelerated in shock waves around the expanding shells of supernovae, and they emit synchrotron radiation when they encounter galactic magnetic fields.

Immediately following Jansky's and Reber's discoveries, the embryonic science of radio astronomy had to be put on hold by the advent of World War II but, afterwards, the tremendous technical advances in electronics, radio, and radar technology that had taken place during the war years enabled the field to take off rapidly. Radio scientists such as James Hay, Bernard Lovell, and Martin Ryle in the UK, and Joseph Pawsey in Australia, all of whom had been involved in the development of radar, turned their attention to the new astronomy.

Large single-dish radio telescopes were constructed, such as Bernard Lovell's 250-foot (76m) diameter steerable Lovell Telescope (Figure 11), built in 1957 at Jodrell Bank Observatory, England. Its many observations include monitoring space probes such as the first artificial satellite, Sputnik, as well as mapping planets, pulsars, galaxies, and quasars. Other single-dish telescopes include the famous 305m-diameter dish of the Arecibo Observatory in Puerto Rico (which collapsed catastrophically in

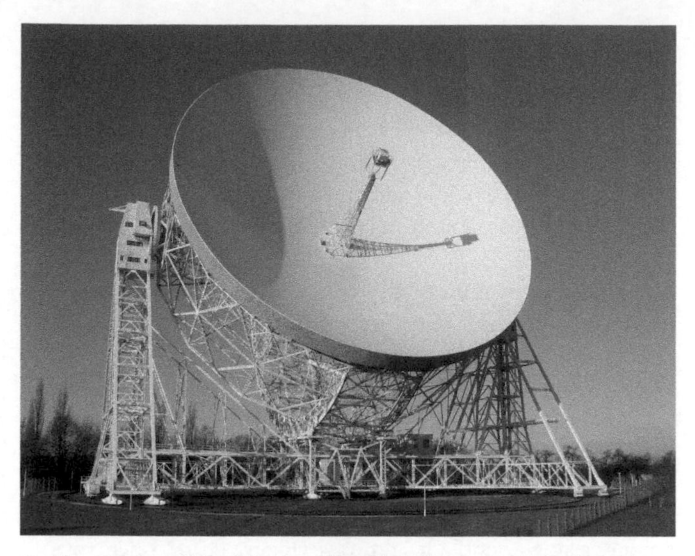

11. The 250-foot dish of the Lovell Radio Telescope at Jodrell Bank.

2020), the steerable 100m-diameter Green Bank Telescope (GBT) in West Virginia and the Five-hundred metre Aperture Spherical radio Telescope (FAST) in Guizhou, China. The GBT is itself a replacement of an earlier telescope which also collapsed catastrophically in 1988, indicating how close to engineering limits large telescope technology is being pushed.

Interferometers

Radio wavelengths are a million times longer than those of visible light. So, for a single-dish radio telescope to match the angular resolution of an optical telescope, its aperture would have to be bigger by the same factor. While it is not possible to make such a large single dish, it is nevertheless a fact that radio telescopes have not only equalled the resolving power of their optical counterparts, but in some cases they have exceeded it. How was this done? Big radio telescopes consist of arrays of many small antennas, often spread over many kilometres, sometimes spanning continents, but always connected together to work as one instrument. An antenna array can be broken down into a number of basic units of which one is a pair of antennas connected together as an *interferometer*. While a single antenna might have a coarse resolution, a pair of antennas, configured as an interferometer, can achieve much higher resolution.

The phenomenon of the interference between two waves was used to prove that light is a wave, and was first demonstrated by the English scientist Thomas Young in 1801. When waves come together, they add to and interfere with one another (Figure 12). By throwing two pebbles into a smooth pond, the circular ripples from each spread out and cross through those from the other. Wherever ripples overlap, each adds itself wholly to the other. Where two equal crests meet, they interfere constructively to produce a single double-height wave (Figure 12(a)), where two equal depth troughs meet they make one of double depth, and where a crest meets a trough they cancel completely (destructive interference) so that the water remains level (Figure 12(b)). Young showed that light has precisely these properties. In his famous double-slit experiment (Figure 12(c)), light is prepared by diffracting it through a narrow slit in a screen, causing it to spread out on the other side. This ensures the light is coherent and comes from the same point-like source. The light then passes through a *pair* of slits in a second screen and again spreads out, but this time it interferes with itself to form a set of bright and dark stripes (such as at points C and D of Figure 12(c)) on a screen; this is a diffraction pattern containing interference fringes. The bright fringes indicate constructive interference and are regions where the paths from the two slits differ by exactly a whole number of wavelengths, so that the crests of the waves arrive there together and add to each other. The dark fringes indicate destructive interference where the wave trains differ by an odd number of *half* wavelengths, arrive out of phase, and cancel each other.

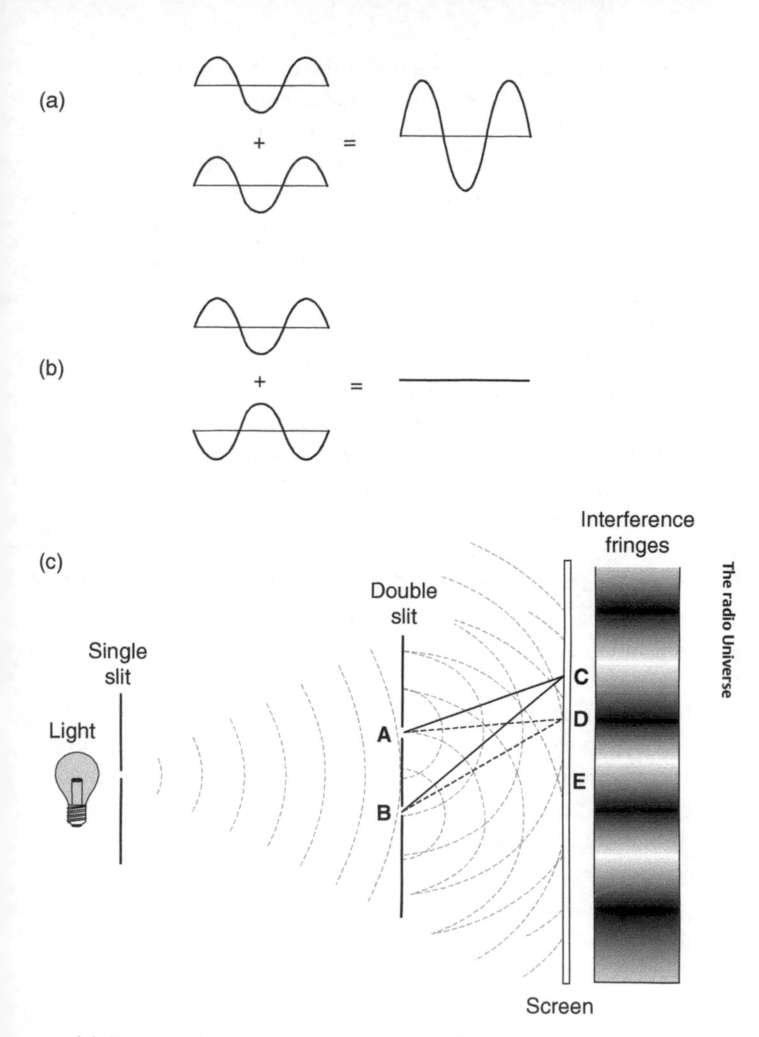

12. (a) Constructive and (b) destructive interference of two waves; (c) Thomas Young's double slit experiment.

Young's experiment translates to a radio interferometer. In Figure 13(a) two radio antennas, A and B, are connected so that their signals add together, forming an interferometer. The antennas are placed on an east–west baseline, a distance D apart, where they are both pointed towards a radio source. An incoming wavefront reaches antenna B first. But, to reach antenna A, the wave has to travel an extra distance AC, the path difference. The time delay between the two signals depends on the direction of the source and the spacing between the antennas. As the Earth rotates, the radio source drifts through the beam of the interferometer, changing the path difference AC and generating interference fringes as the two signals interfere with each other. The fringes show up as a sinusoidal variation of the interferometer output which is shown in the figure as modulating the much broader envelope of a single-antenna beam pattern. The fringe pattern can be thought of as a series of lobes projected onto the sky, spaced apart by an angle of the ratio of the wavelength to the antenna separation, D. This angle defines the angular resolution of the interferometer.

The difference in brightness between the maxima and minima of the fringes is called the fringe visibility, an important quantity that contains information about the angular distribution of the source, i.e. its radio 'image'. There are three possibilities. First, if the source is point-like, such as a quasar, the fringe visibility is large and, in between interference peaks, the interferometer output falls to zero indicating that the source is not resolved. Second, if the source is broader and therefore partially resolved, the minima do not reach zero, and the fringe visibility is reduced. Third, if the source is uniformly bright across the sky, the output is constant and no fringes are produced.

Comparing Young's experiment with the radio interferometer, one can think of the light from the single slit as representing the point radio source, the two slits in the second screen representing the pair of radio antennas, and the fringes projected on the screen as

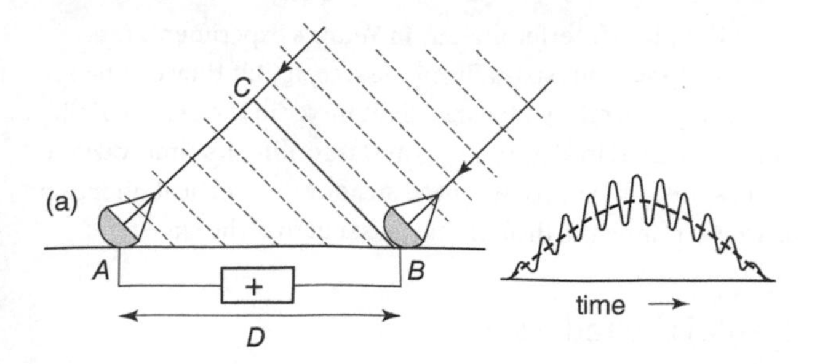

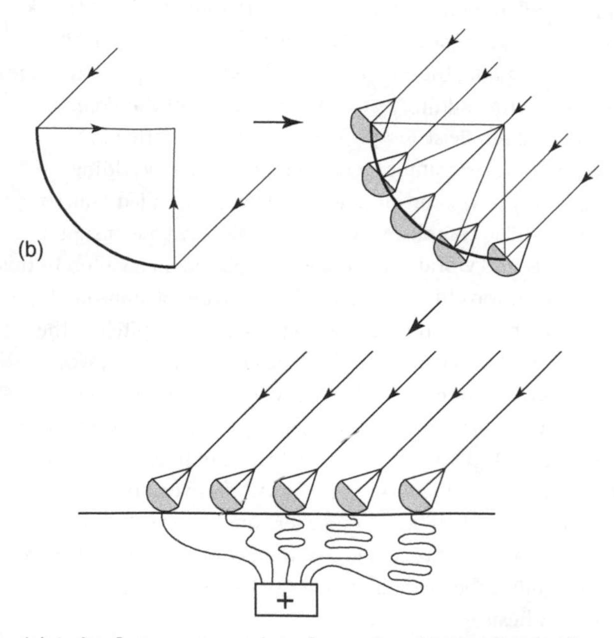

13. (a) A simple two-antenna interferometer with baseline *D*. The two antennas are connected to a receiver via cables, where they are added (+) to produce the time-trace shown on the right. (b) Synthesizing a giant dish antenna. (Right) the surface of the dish is divided into a number of areas containing small antennas. (Lower) the small antennas are now moved to ground level where they are connected to a receiver and, added via cables of different lengths, adjusted to simulate the various path time delays of different parts of the big dish.

e of the radio interferometer. In Young's experiment, the nges are spread out spatially on the screen. But they can be made to vary in time by placing a light meter in the centre of the screen (at point E in Figure 12(c)) and recording its time-varying output as the light source is moved steadily in a vertical direction. This would mimic the drift of the quasar across the sky.

Mapping the radio sky

To turn interferometer measurements into radio maps of the sky, one more idea is needed. With the example of a pair of pebbles thrown into a pond, the complex wave interference pattern in the water comes from adding just two waves. In 1822 the French mathematical physicist Joseph Fourier discovered that *any* waveform, however complex, can be represented by adding together many pure sinusoidal waves. These are called Fourier components. The complex waveform can be any type of signal, such as the sound vibrations from a note played on a violin. In this case the sound waveform consists of a spectrum of sinusoidal tones based on a fundamental note (the perceived pitch of the note), plus a set of harmonics, the Fourier components, which add together to produce the familiar violin timbre. Fourier's method is a powerful way of representing all kinds of waveforms, including the 2D distribution of brightness in images, such as a photograph or a radio map of the sky. All these can be decomposed into their spatial Fourier components. Provided enough components are added together, considering their phases and amplitudes, the original image can be reconstructed using Fourier synthesis.

Aperture synthesis is the technique of creating one very large telescope aperture from a number of interferometers. A single-spacing interferometer is sensitive to one spatial component of the sky brightness distribution and, by varying the spacing, the interferometer produces a set of Fourier components, corresponding to structure on different angular scales. Imagine dividing up the surface of the very large radio telescope aperture to be synthesized into many smaller areas, each containing a smaller antenna (Figure 13(b)). The signals from all the antenna pairs are connected and added with the correct phases so as to mimic the time delays of each of the rays reflecting from the big dish. The time delays can be introduced either by using different length interconnecting cables or optical fibres (as shown) or, as is now done, by correlating the signals in a computer. In practice it is unnecessary to cover the *whole* area of the imagined large dish with small antennas, nor is it essential to acquire all the signals simultaneously provided the source does not vary during the observations. Instead, parts of the dish can be sampled at different times, and a smaller number of antennas moved around to build up the radio image gradually.

An important development in radio astronomy was *Earth-rotation synthesis*, in which antennas are arranged on an east–west line. Owing to the 24-hour rotation of the Earth, the antennas effectively rotate around each other every day, as seen from the distant stars. By repeating the observations with different antenna spacings, it is possible to sample the different Fourier components of the source. Earth-rotation synthesis was pioneered by English astronomer Martin Ryle at the Mullard Radio Astronomy Observatory (MRAO) in Cambridge, England, using three 60-feet-diameter dishes of the One-Mile Telescope riding on an east–west section of the long-defunct Oxford to Cambridge railway line. The first aperture synthesis radio map, the North Pole Survey, was published in 1962. A large modern aperture synthesis array is the JVLA telescope (Figure 14), consisting of 27 25m dishes, laid out in a Y-pattern with 21km-long arms. The dishes can be moved around on railway tracks. The telescope operates at wavelengths from 30cm to 6mm, at which the angular resolution is 0.04 arcseconds, comparable with the resolving power of a large optical telescope.

14. The Karl G. Jansky Very Large Array radio telescope in New Mexico.

Quasars

From the beginning, interferometer observations revealed many bright, discrete radio sources but, at first, the measured positions were too imprecise to identify optical counterparts. As techniques improved, two of the strongest radio emitters, Cygnus A (first observed by Reber in 1939) and Cassiopeia A, were identified respectively with a young supernova remnant in the Milky Way, and a faint galaxy with a redshift distance of 760 million light years. The big surprise was the discovery that the radio emission from Cygnus A came not from the body of the galaxy, but from two enormous radio lobes dwarfing the central optical galaxy (Figure 15). The synchrotron-emitting radio lobes in Cygnus A are powered by two highly aligned beams of particles and radiation (jets) emerging from the central galaxy in opposite directions, which are highly collimated over enormous distances. In the

15. The JVLA radio image of galaxy Cygnus A. The double radio source is about 540,000 light years across and its two vast radio lobes are powered by narrow jets protruding in opposite directions from the much smaller central galaxy. The jets transport beams of particles and radiation which plough into the intergalactic medium dispersing the energy.

centre, the AGN harbours a powerful quasar, a supermassive black hole which is actively accreting matter.

Many of the bright radio sources seen in the first large-scale radio surveys could not be resolved, even with the widest interferometer spacings then available. The position of one of the brightest of these sources, 3C273 (object 273 in the 3C or the Third Cambridge Catalogue of Radio Sources), was pinpointed by Australian astronomers in 1963 using the 64m Parkes radio dish using the technique of lunar occultation. The positional information enabled Dutch astronomer Maarten Schmidt to use the 200-inch Palomar telescope to identify 3C273 with a peculiar bright blue star-like galaxy, emitting continuum radiation and having unusually broad optical emission lines indicating highly excited gas. This, the first quasar, was strongly redshifted ($z = 0.158$) implying a distance of three billion light years. This distance enabled astronomers to infer that the galaxy has an unprecedented luminosity of four trillion times that of the Sun, roughly equivalent to the light output of about 100 Milky Way axies. More quasars were soon identified, many with even higher redshifts. As these data accumulated, the number density of quasars could be studied over cosmological timescales and distances. The statistics revealed that the quasar population peaked at redshifts between $z = 1$ and $z = 3$, corresponding to light-travel times between about six and eight billion years. Quasar observations were crucial in showing that galaxies have undergone significant evolution in the history of the Universe.

Supermassive black holes and their jets

Once it had been established that quasars are the sites of the most powerful sustained energy releases known in the Universe, the question was: what could possibly light up one galaxy with the power that illuminates a hundred? One clue came from the observation of their rapid time variability. The luminosities of quasars can fluctuate on short timescales, as rapidly as in tens of minutes, indicating that their emitting regions are extremely small, less than a few AU. (Any object whose luminosity varies rapidly must be physically smaller than the light-travel time across the object, otherwise the variations are smoothed out.) The most powerful energy release mechanism then known was the process that powers the stars, thermonuclear fusion. However, fusion reactions can release at most only around 0.7 per cent of the rest mass energy of matter, which was considered insufficient to explain the luminosity of a quasar.

In 1963 the New Zealand mathematician Roy Kerr solved Einstein's equations for a spinning black hole which showed that the spacetime outside the event horizon, and any objects that happen to be there, get dragged around the hole as if caught in a whirlwind. This prompted the English physicist Roger Penrose to propose that energy can be extracted from a spinning black hole by the fragmentation of objects falling into the black hole from outside the event horizon; some fragments fall in, and others fly off. Depending on the details, up to 42 per cent of the rest mass energy of matter can be extracted from a spinning black hole, a much more efficient process than fusion and one that is now the leading candidate for a quasar's power source. The great luminosities of quasars therefore pointed to the release of gravitational binding energy when matter accretes onto a supermassive black hole. The strong curvature of spacetime surrounding a black hole is indicated in Figure 16(a), showing the deep gravitational potential energy well. An object falling towards a black hole from infinity acquires a large amount of kinetic energy from the gravitational field. Think of the water in a gently flowing river gradually picking up energy from the Earth's gravitational field as it approaches a waterfall before cascading downwards at high speed.

In an active quasar, such as the one depicted in Figure 16(b), radiation exerts an outward pressure force on the infalling matter. For accretion onto any massive luminous body there is a maximum luminosity (the 'Eddington luminosity') above which the radiation pressure will overwhelm gravity and blow away any material which is trying to accrete. English astrophysicist Arthur Eddington had originally introduced this limit when considering the accretion of matter onto massive stars. When the Eddington

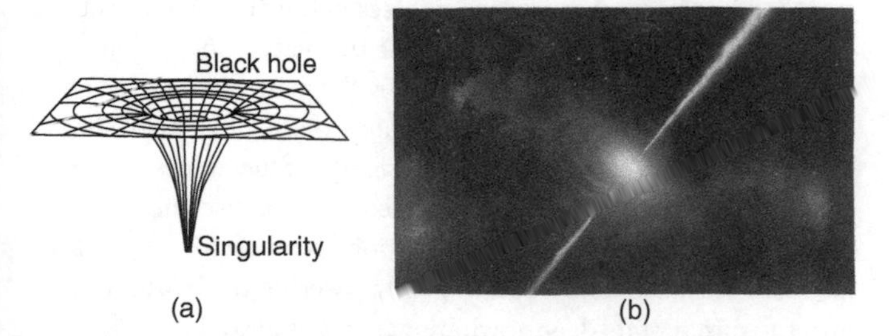

16. (a) The strongly curved spacetime and gravitational potential well around a black hole. (b) Artist's impression of one of the most distant quasars known (P172+18), about 13 billion light years away. It shows the accretion disc around the supermassive black hole, and the highly energetic jets of matter and energy which produce radio emission.

limit was applied to the extreme luminosities in quasars like 3C273, it showed that the central object must have a mass of at least 1 billion solar masses to hold it together. A supermassive black hole with this mass has an event horizon of three billion kilometres; on the scale of the solar system such a black hole would fit comfortably within the orbit of Neptune.

The infalling matter powering an AGN can be in any form, for example, an interstellar gas cloud, star, or planet. As it approaches the black hole, the material experiences strong tidal forces which stretch it out into long strands, a process that has been called *spaghettification*. The matter will generally possess some rotational angular momentum, causing the material to flatten out into a thin, dense *accretion disc* rotating around the black hole. As the matter falls in, it spins up, in line with the law of conservation of angular momentum. A pirouetting ice skater uses the law when she first goes into a slow spin with arms outstretched and pulls them in to her body to produce a spectacular fast spin. The viscous friction between the fast-spinning inner disc and the slower rotating outer material heats the material to tens of millions of degrees, where it radiates copious amounts of gamma rays and X-rays. Further out, in the cooler parts of the disc, the radiation peaks respectively in the ultraviolet, optical, and infrared wavebands. The viscous frictional forces in the disc also transfer angular momentum outwards, causing the matter nearest the black hole to spiral inwards. Eventually the matter arrives at the last stable circular orbit, which for a non-rotating black hole has a radius of 3 Schwarzschild radii (and smaller if the hole is rotating). There, it falls off the inside edge of the disc and passes through the event horizon. The accretion disc therefore regulates how fast the black hole feeds, by gently lowering the matter down into the throat of its deep gravitational potential well.

Active supermassive black holes in galaxies, their accretion discs, and their jets are the cosmic particle accelerators that power quasars. While the details of their workings are not well

understood, there are some basic, well-accepted ideas.
The material in the accretion disc is hot ionized plasma, threaded by magnetic field lines which rotate around with the disc. It is likely that rotation winds up the magnetic field like spaghetti being wound around a fork, collimating the emerging beams of matter into narrow jets. The huge electric fields that exist in this region accelerate particles to ultra-relativistic energies, and shoot them out in both directions along the rotation axis. The spin of the black hole acts like a gyroscope, keeping the jets aligned to one direction. Further out, the jets punch their way out of the host galaxy, and plough into the tenuous intergalactic medium where, at great distances, they deposit their energy.

Observationally, there are two main classes of AGN: *radio loud* and *radio quiet* types. If the central black hole is feeding voraciously on local sources of matter, it produces the powerful jets and radio lobes of the radio loud variety. If the black hole cannot access much matter, there is no significant emission from jets, the object is radio quiet, and lies dormant. AGN produce a broad range of observational phenomena depending in part on the observed orientation of the accretion disc. Encircling the accretion disc, and typically a few light years across, is an outer hollow doughnut-shaped ring (torus shape) of hot dusty material which absorbs the high-energy radiation from the centre. A range of different observational properties is produced depending on the orientation of the accretion disc with respect to the observer's line of sight, as well as on the feeding activity of the black hole. For example, if the AGN is hidden by dust, the central radiation is absorbed and reradiated by the dust. In this case the AGN produces characteristic thermally broadened spectral lines. If, on the other hand, an active jet happens to be pointing almost towards the observer, it appears exceptionally bright and is known as a *blazar*. The increased brightening is caused by the enhancement of radiation emitted fast particles in the forward direction of motion, called relativistic beaming. Supermassive black holes in AGN are also believed to regulate the rate of growth galaxies. In an active phase, the jets deliver kinetic energy to the surrounding gas and, by heating it, prevents it from condensing down onto the galaxy.

Gravitational lensing

The simplest geometrical optics model of light tells us that it travels along straight lines in a vacuum. This works well on small length scales and away from any large masses. However, Einstein's General Theory of Relativity predicts that light travels in straight lines only if the Universe is empty of matter. When mass is present, spacetime is curved, and light follows a curved (or geodesic) path. The American relativist John Wheeler put it crisply: 'matter tells space how to curve, and space tells matter how to move'. Einstein's prediction was confirmed in 1919 when Eddington measured the deflection of starlight around the Sun during a total solar eclipse, a special time when it is possible to observe distant stars close to the limb of the Sun.

Einstein had originally considered the possibility of observing the light from a distant star being deflected (lensed) by the gravity of a closer one that happens to be lying along the line of sight. The probability of such an alignment was, however, considered negligibly small. But the probabilities of lensing alignments are much bigger in the case of galaxies and clusters of galaxies. The first *gravitational lens* was observed in 1979 in observations of the 'twin quasar' (Q0957+561), two closely spaced quasars that appeared to have similar spectral signatures. In fact, it turned out that this odd object was a *single* quasar whose light rays had passed along two paths near an intervening mass, as indicated in Figure 17.

The enormous gravity of a cluster of galaxies can act as a lens to focus the light emitted by background objects, behaving like the objective lens of a telescope positioned out in deep space. The lens distorts the shapes of background objects by magnifying them and

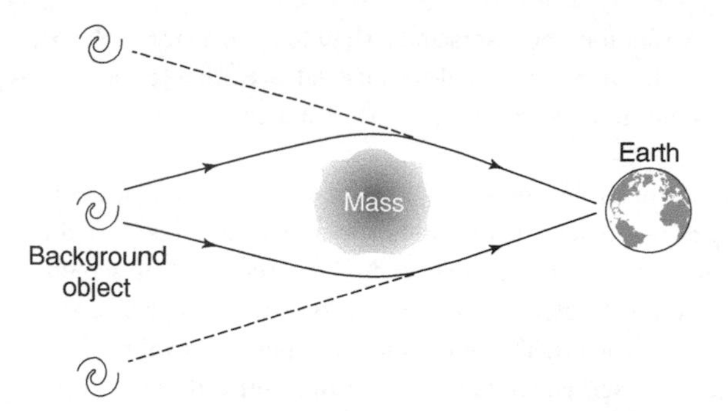

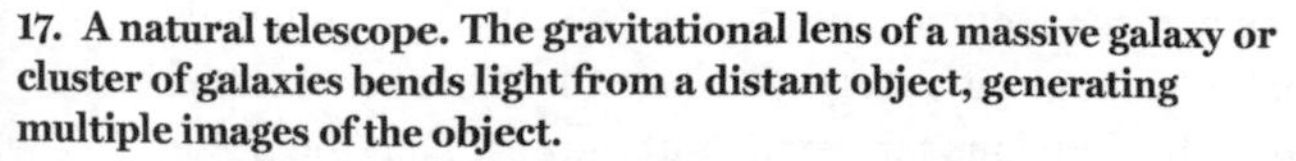

17. A natural telescope. The gravitational lens of a massive galaxy or cluster of galaxies bends light from a distant object, generating multiple images of the object.

shearing the light paths (tangentially stretching the image around the lensing mass). With *strong lensing*, distant galaxies appear as curved arcs or sometimes even full circles (*Einstein rings*) around the lensing masses.

Pulsars

In the 1960s, English radio astronomer Anthony Hewish built a special type of radio telescope, the Interplanetary Scintillation Array (ISA), at the MRAO in Cambridge, UK, which was designed to discover quasars. It was based on the principle that, at radio wavelengths, the bright point-like quasars are observed to 'twinkle' (or scintillate) owing to their ray paths being multiply refracted in the turbulence of intervening interplanetary and interstellar plasmas. This effect is analogous to the 'twinkling' of the stars due to the passage of their light through the Earth's atmosphere. Optically, planets do not appear to twinkle simply because, with their larger angular sizes, there is more cancellation of the fluctuations along multiple refraction paths. So, by observing how much a quasar scintillates, the ISA telescope could be used to estimate its angular diameter. The key feature of this telescope is that it needed to have a rapid response time to detect the scintillation of quasars and so had to have a large collecting area to maximize the signal-to-noise ratio. Such a telescope design turned out to have unexpected consequences.

The ISA occupied an area of about two and a half football pitches and consisted of thousands of wire dipoles, tuned to a wavelength of 3.7m, and backed up by wire reflecting screens. It could be directed to different parts of the sky by electrically phasing the array. The telescope was operated initially by Hewish's research student Jocelyn Bell Burnell, who, in 1967, came across a strange radio source (CP1919) pulsing with clock-like regularity, every 1.33 seconds. This could have been caused by mundane terrestrial interference, but the pulses always came from the same part of the sky. When other similar pulsing sources were found, with different periods, interference was ruled out. This was how pulsating radio stars, or *pulsars*, were discovered.

What kind of cosmic radio source produces regular ticks like a clock? The most likely explanation was that it must be some kind of small rotating object—sending out flashes like a lighthouse. The first thought was that the pulsar might be a *white dwarf*, the core of a star weighing less than 8 solar masses that, having run out of fusion fuel, has collapsed to a dense hot ember, about the size of the Earth. The pressure that stabilizes a white dwarf against gravitational collapse is electron degeneracy pressure. Electron degeneracy pressure stems from the quantum mechanical properties of electrons which prevents electrons being squeezed together too tightly. Electrons obey the Pauli exclusion principle which forbids any two electrons in an atom from occupying the same quantum state and is named after Austrian physicist Wolfgang Pauli. The Indian astrophysicist Subrahmanyan Chandrasekhar calculated that a star weighing up to about 1.4 solar masses (the Chandrasekhar limit), but no more, can be supported by electron degeneracy pressure alone. A white dwarf is one of the densest known forms of matter, exceeded only by neutron stars and black holes. (A white dwarf is about the size of the Earth, but its density is around 200,000 times higher.)

It turns out that even a star as dense and compact as a white dwarf could not withstand the rapid rotation observed in pulsars—it would quite simply fly apart under the centrifugal forces. The observations instead pointed to an object that is bound together much more strongly, a neutron star. As early as 1934 astronomers Fritz Zwicky and Walter Baade had predicted the possible existence of neutron stars, as being produced in core-collapse supernovae, but up to this point they had never been observed. The quantum degeneracy pressure supporting a neutron star against gravity comes from the *neutrons*. A neutron star can be thought of as being akin to a giant atomic nucleus, a tiny spherical object only around 10km across but weighing around 2 solar masses, with a density about a million times higher than a white dwarf. A famous example is the pulsar in the Crab Nebula (Figure 18, top left image), which flashes 30 times a second. Other pulsars are observed to have periods as short as a millisecond.

A complete description of the underlying radiation mechanism of a pulsar is a complex and active research area, but the broad picture is this. A pulsar is believed to be a spinning, magnetized neutron star, the product of a core-collapse supernova. As in the pirouetting ice skater example, the conservation of angular momentum means that any small initial rotation of a star is amplified many times over during the collapse. A tiny, rapidly rotating object weighing perhaps 2 solar masses has a huge inertia which means that its rotation period will retain its clock-like precision over long periods of time. The signals from some pulsars, for example the Vela pulsar, have been observed to show sudden jumps (glitches) in the pulse rate. These are believed to be connected with 'starquakes', sudden contractions of the surface crust of the neutron star by distances in the order of centimetres.

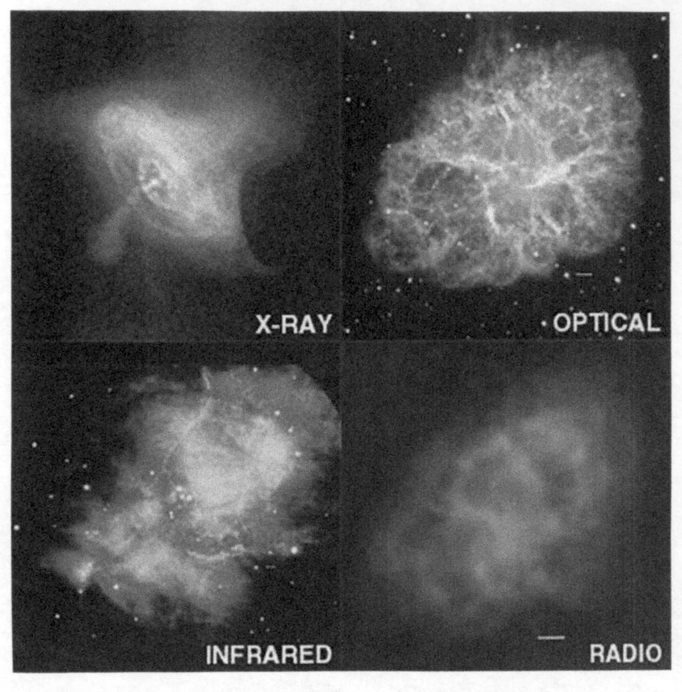

18. The Crab Nebula, shown at different wavelengths, is the remnant of the supernova explosion of a massive star that exploded in AD 1054, and contains a pulsar in the centre.

The progenitor star will, prior to the explosion, have been threaded by a magnetic field which, during the collapse, is amplified enormously to make a magnet about a billion times stronger than a fridge magnet. A magnetized, rapidly rotating neutron star is a powerful electric generator. The electric field at the surface of the star is strong enough to pluck electrons out and accelerate them to near-light speed where, deflected into helices by the magnetic field, they emit synchrotron radiation. More than 200 pulsars have now been discovered and they are observed to radiate pulses across all wavebands.

Another telescope with no moving parts

The development of fibre optic cables and massively powerful supercomputers has opened up efficient ways of building radio telescope arrays based on many omnidirectional fixed antennas spread over the surface of the Earth. These arrays employ 'beam-forming' phased-array methods enabling them to be pointed anywhere in the sky. This approach resulted in 2012 in a Dutch telescope, the LOw Frequency ARray (LOFAR), which operates at the longest wavelengths (2m–10m) observable from the Earth. The individual antennas are cheap, mass-producible items distributed over distances of hundreds of kilometres at sites from Ireland in the west to Latvia in the east, and from Sweden in the north to Italy in the south. LOFAR's antennas can look in many directions at the same time. Being able to see a large part of the sky makes LOFAR a powerful survey telescope and therefore one well suited to discovering transient sources. Astronomers are hoping that LOFAR's low frequency capabilities may shed light on one of the most important questions in astronomy: what happened in the *Dark Age* of the Universe. The Dark Age is a time before the first stars had formed when all normal matter was in the form of clouds of hydrogen and helium. The radio emission and absorption at a wavelength of 21cm from neutral hydrogen atoms from this epoch is thought to be observable but, owing to the expansion of the Universe, has been redshifted to wavelengths of around 4m. LOFAR hopes to observe this radiation which will be critical to our understanding of the earliest times immediately after the Big Bang.

Probing the cold Universe

Hydrogen is the most abundant constituent of the Universe, existing as ions (protons), neutral atoms, and H_2 molecules. The ground state of a hydrogen atom has two closely spaced (hyperfine) energy levels, and atoms flipping quantum states between the parallel/anti-parallel spins of the electron and proton produce a radio (HI) line at a wavelength of 21cm. Observing the 21cm radio line, both in emission and absorption, is an important branch of astronomy. Spectroscopic observations of the HI line emission intensities and Doppler shifts yield information on the quantity of hydrogen and its kinematics in galaxies. Since HI radiation is not appreciably absorbed by interstellar dust, it is possible to use it to observe the Milky Way's hydrogen across the full width of the galactic disc, and this has revealed the otherwise unobservable spiral structure of our galaxy. Other atoms and molecules emit and absorb line radiation at radio and far-infrared wavelengths. The first molecule detected in the interstellar medium was the hydroxyl radical, OH, which has shown remarkable maser-like amplification properties in dense regions of the interstellar medium. (A maser is the microwave analogue of a laser.)

Spectral lines of other complex molecules lie in the submillimetre wavebands. Submillimetre telescopes require mirrors with near-optical surface accuracies and employ cooled germanium crystal bolometer detectors to observe the total radiant power. These instruments must be cooled to within a few degrees of absolute zero. Observational astronomy at such short radio wavelengths is challenging: astronomers have to work within the narrow windows between the many atmospheric water vapour absorption lines and only a handful of sites in the world are dry and dark enough for these observations. One is the summit of the dormant volcano of Mauna Kea in Hawaii, the site of the 15m James Clerk Maxwell Telescope (JCMT), the largest single-dish telescope in the world operating in the 0.3mm–2mm waveband. It lies above a third of the Earth's atmosphere and most of its water vapour. The JCMT has observed the emission from a population of *starburst galaxies* at large redshifts, a type undergoing extremely rapid star formation in which the intense ultraviolet radiation from hot newly formed stars is reradiated by dust at far infrared and submillimetre wavelengths. Observations show that the star-forming rate in galaxies peaked about 10 billion years ago, about 2–3 billion years after the Big Bang, and has now dwindled to a much lower rate.

The submillimetre and millimetre wavebands contain a plethora of spectral lines from relatively complex molecules, including water, ammonia, ethanol, and methanol. While molecules can form in the interstellar medium by association in the gas phase, formation is more efficient when molecules stick to the surfaces of dust grains, conditions that prevail in star-forming regions. These stellar nurseries are associated with the million-solar-mass Giant Molecular Clouds (GMCs), dense regions of the interstellar medium where temperatures are just a few tens of degrees above absolute zero, and gas clouds readily clump together under gravity. GMCs shine brightly in the far-infrared and are observed with telescopes equipped with arrays of highly sensitive superconducting bolometric sensors.

The most powerful telescope in the world studying the cold Universe is the Atacama Large Millimetre/submillimetre Array (ALMA), an aperture synthesis telescope operating at wavelengths between 0.3mm and 3.6mm, sited in one of the driest places on Earth—the Chajnantor plateau, 5km high in the Chilean Andes (Figure 19). ALMA has 66 paraboloidal dishes, linked by fibre optic cables. The dishes are moveable, and can either be gathered together to give a wide-angle view, or spread out 16km apart to zoom in on fine detail with an angular resolution of 0.006 arcseconds. Making images at these short wavelengths requires high precision—each antenna must be located on the ground to a tolerance of about 25 microns (one-third of the width of a human hair). One of ALMA's spectacular images is of the protoplanetary disc system around the young star HL Tauri (Figure 20), showing planets forming in a disc. In other protoplanetary discs, ALMA has identified carbon-based molecules, such as methyl and hydrogen cyanide and the spectral signatures of complex organic molecules, the building blocks of life.

19. Antennas of the Atacama Large Millimetre/submillimetre Array, 5km high in the Chilean Andes.

20. HL Tauri: a solar system in the making, a young star and its protoplanetary system observed with the ALMA telescope. The star is surrounded by a disc of gas and dust, dark rings swept clean by newly emerging planets. The disc is about twice the size of the solar system.

Very long baseline interferometry

The high angular resolution achievable with short-wavelength arrays like ALMA comes from distributing antennas over distances of tens of kilometres. It is possible to increase the resolving power of ground-based radio telescopes even further by extending baselines to hundreds of kilometres, across continents, and ultimately distances limited by the diameter of the Earth. But to do this the signals from the antennas must be correlated together accurately. By the late 1960s, the technique of *Very Long Baseline Interferometry* (VLBI) was first demonstrated when radio dishes located thousands of kilometres apart were synchronized to create a telescope having an angular resolution of thousandths of an arcsecond. Since it is physically impossible to use cables or fibres to interconnect such far-flung antennas, the signals at each are recorded digitally, time stamped using atomic clocks, and correlated at a later time.

VLBI is used to observe bright sources on the finest angular scales, such as extragalactic radio sources and planetary surfaces. One of the great discoveries of VLBI, of the molecular water maser line in the AGN of galaxy M106 at a wavelength of 1.3cm, was made in 1995 with the USA's Very Long Baseline Array (VLBA). Because a maser emits over an extremely narrow range of frequencies, the velocities of water masers orbiting a supermassive black hole were measured by radio spectroscopy right into the centre of the galaxy. These clearly showed the Keplerian orbital rotation of the material from which it was possible to infer the presence of a 40-million-solar-mass supermassive black hole.

The shadow of a black hole

In 2017, a 200-strong team of astronomers of the Event Horizon Telescope (EHT) used a network of eight telescopes from all over the world to make the first ever image of the shadow of a black hole. Black holes are small, dark, and distant, properties that present major observational challenges. The technique focused on observing the *shadow* of a supermassive black hole in an AGN, silhouetted against the bright background of the accretion disc. In order to view the structure close to the event horizon, observations had to be made at a wavelength at which the plasma is transparent. This and the need for high angular resolution could both be satisfied by observing the black hole at a wavelength of 1.3mm using an Earth-sized VLBI telescope. The millimetre-wave observatories involved were: ALMA and the Atacama Pathfinder Experiment in Chile, the Spanish 30m Institute for Radio Astronomy in the Millimetre Range telescope, the Large Millimetre Telescope in Mexico, the Submillimetre Telescope in Arizona, the JCMT and the Submillimetre Array telescopes in Hawaii, and the South Pole Telescope in Antarctica. The supermassive black hole selected for observation is in the heart of the giant elliptical galaxy M87 and weighs 6 billion solar masses. To image it required a telescope with an angular resolution of 0.000001 arcsecond. A telescope the size of the Earth was needed to see an object the size of an apple on the surface of the Moon.

The EHT image (Figure 21) shows a bright asymmetrical ring surrounding a dark centre. What are we actually seeing in this image? The ring of light is emitted by the hot million-degree plasma in the accretion disc which is swirling around the black hole at an appreciable fraction of the speed of light. The accretion disc does not reach all the way down to the event horizon because there are no stable circular orbits less than 3 Schwarzschild radii around a non-spinning black hole. Matter that crosses inside the last stable orbit falls into the black hole and disappears forever. Outside the event horizon of the black hole is a surface called the *photon sphere*, which has a radius of 1.5 Schwarzschild radii (Figure 22). Space is so strongly curved at the photon sphere that photons are able to follow closed orbits around the hole. An imaginary observer sitting on the photon sphere could look out along it and see the photons emitted from the back of her head!

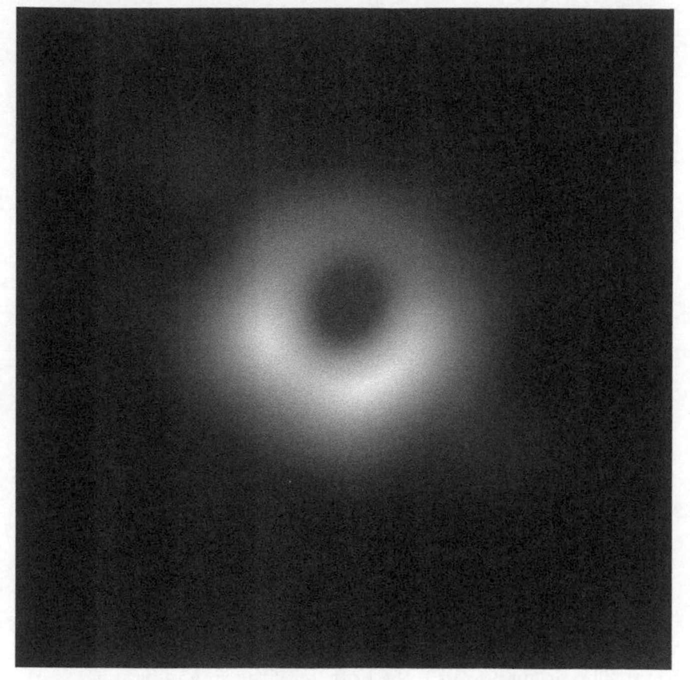

21. The image of the shadow of the supermassive black hole in the centre of galaxy M87.

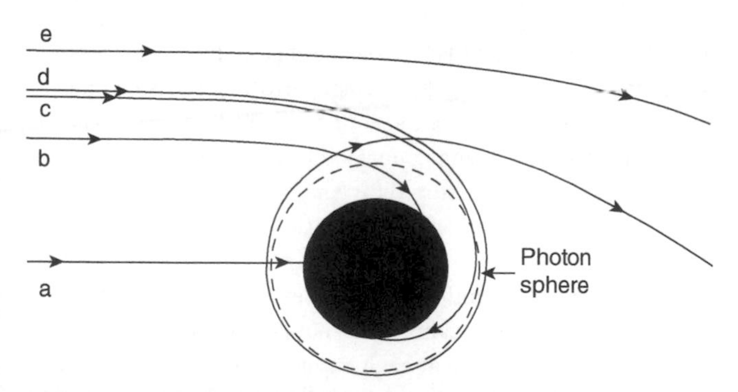

22. Light ray paths around the event horizon of a black hole (filled circle). Rays a, b, and c cross the photon sphere and are absorbed by the black hole. Ray d just grazes the photon sphere and can escape having made one circuit.

The photon sphere separates the paths of photons that either spiral into the black hole (orbits a, b, and c), or escape (orbits d and e). Rays crossing the event horizon will disappear into the black hole. If a light ray approaches just above the horizon, it will be bent towards the black hole, cross the event horizon, and also disappear. Even a ray approaching at the radius of the photon sphere will suffer a similar fate. It turns out that for an incoming parallel ray to graze but not cross the event horizon, it must approach at more than 2.6 Schwarzschild radii; this is the radius of the shadow of the black hole.

The innermost region of the observed dark shadow in Figure 21 is the event horizon facing the observer, and the outer part of the shadow maps the *back* of the event horizon. This strange situation arises because of the very strong warping of space around the black hole which allows us to see both front *and* back surfaces in one image. Further out in the image, the light from the inner edge of the bright ring comes from rays that just graze the photon sphere and can escape to reach us. The accretion disc in M87 happens to be orientated almost perpendicularly to our line of sight. The bright region at the bottom of the ring of light is relativistically beamed, indicating that the plasma on this side is approaching us.

To form this image, the EHT telescopes had to be synchronized precisely at each of the eight sites using hydrogen maser atomic clocks, accurate to within one second in ten million years. The enormous quantity of data generated by the telescopes (about 350 terabytes a day) was stored on computer hard drives and brought physically together for correlation and processing. Reconstructing the image from the data was not straightforward. Unlike well-sampled synthesis radio telescope data, these VLBI measurements are sparse and it was not possible to obtain absolute phase calibration at each observatory. A sophisticated algorithm was developed to correlate the data, filter out noise and distorting atmospheric effects, and find the best-fitting image.

In 2022, the EHT released a second image of the shadow of the supermassive black hole in the centre of the Milky Way, close to the bright radio source Sgr A*. This image is broadly similar to the one in M87. However, this observation proved to be more challenging to make because even though Sgr A* is closer to us, its mass is 1,000 times smaller, it is currently not accreting much matter (making it less luminous), and the line of sight to it passes through the Milky Way's interstellar medium which tends to scatter the radio waves. Also, being a smaller black hole, the hot gas swirling around it at relativistic speeds resulted in significant signal fluctuations during the 8–10 hours observation time.

The Square Kilometre Array

The Square Kilometre Array (SKA) is a radio telescope with a collecting area of literally 1 square kilometre (i.e. 1 million square metres), which will be composed of thousands of radio dishes spread across two continents, from the Karoo region of South Africa to the Murchison region of Western Australia. It will also have up to a million low frequency antennas. Thirty years in the making, the SKA will be the world's largest scientific instrument, expanding the sensitivity, resolution, and spectral coverage of existing radio telescopes by over an order of magnitude and one that will transform our view of the Universe. The wavelength coverage is wide: from 1cm to 4m. The project is proceeding in steps with the first precursor arrays seeing first light in 2024. One of the first, the 64-dish Southern African *MeerKAT* array spread over 8km, is already producing striking data showing jets from stellar mass and supermassive black holes. MeerKAT has surveyed a 6.5-square-degree region around the supermassive black hole in the Milky Way, Sgr A* (Figure 23). This image reveals the galactic centre to be a complex and chaotic region, with many supernova remnants, clusters of young, massive stars, a high density of cosmic rays, and a tangled web of around 1,000 wispy radio filaments projecting away from the centre of the galaxy. The image

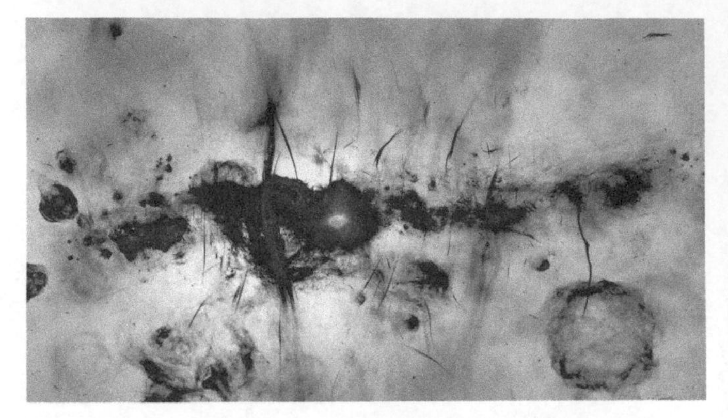

23. MeerKAT radio map of the galactic centre region around the supermassive black hole Sgr A* (in the centre) in the Milky Way.

shows evidence that an explosive event has taken place in the region around Sgr A* a few million years ago.

This chapter has discussed how, by observing in the radio and far-infrared wavebands, we have discovered completely new facets of the Universe. Next, we look at how the rest of the electromagnetic spectrum is opening up by making observations from space.

Chapter 4
Telescopes in space

Space telescopes are needed to observe the large swathes of the electromagnetic spectrum which are absorbed by the atmosphere and cannot reach the ground. These are the mid- and far-infrared, gamma ray, X-ray, and UV wavelengths. By opening up the wider electromagnetic spectrum, we are discovering a Universe teeming with high-energy activity revealing gamma-ray bursts, X-rays from hot gas in clusters of galaxies, the accretion discs around black holes, and UV from massive young stars. Space telescopes show us starburst galaxies shrouded in thick dust cocoons, complex star-forming regions, and extrasolar planets and their atmospheres. We are now seeing the youngest and most distant galaxies that have ever been observed.

Optical and infrared space telescopes

In addition to observing radiation which cannot reach the ground, space telescopes also offer advantages at visible wavelengths. Space telescope images are free of both the turbulent distortions imparted by Earth's atmosphere (enabling sharp diffraction-limited images to be made), and of the auroral light and airglow emission that limit the sensitivities of ground-based telescopes. These advantages extend to the infrared waveband, where thermal heat radiation from the body of the atmosphere itself becomes a limiting factor. In space, infrared telescopes are cooled to avoid the sensors being swamped by thermal noise. For this, some telescopes carry reservoirs of cryogens such as liquid helium to cool detectors to a few degrees above absolute zero to observe at the longest wavelengths. Carrying a finite supply of coolant, however, limits mission lifetimes. The James Webb Space Telescope employs a passive sunshield to block out the Sun's heat, and a cryocooler (a type of low-temperature refrigerator) to cool detectors.

The Hubble Space Telescope (HST; or 'Hubble') was the world's first optical space telescope. The National Aeronautics and Space Administration (NASA), in partnership with the European Space Agency (ESA), launched the 2.4m HST from the Space Shuttle Discovery in 1990 into a low (570km) Earth orbit, where it operates in the near-infrared to the near-ultraviolet wavebands and can produce sharp diffraction-limited (0.05-arcsecond resolution) images. Hubble is famous for its breathtaking images of objects such as the star-forming region the 'Pillars of Creation' in the Eagle Nebula (Figure 24). These towering shapes of interstellar gas and dust have been etched into fantastic forms by the harsh ultraviolet light and fierce stellar winds from hot young stars. This is only one of the many well-known and ground-breaking observations made by Hubble; there are many others which include the surfaces of planets, the protoplanetary discs in star-forming regions which are ejecting jets of material, exoplanets, AGN, the optical counterparts of gamma-ray bursts and supernovae, and measurements of the cosmic web of dark matter and energy.

The HST's first role, however, was to measure the Hubble constant H_0 with much higher precision than had been done before. H_0 is a fundamental cosmological parameter which measures the local expansion rate of the Universe, and determining this important constant depends on making precise astronomical distance measurements. Edwin Hubble's original discovery of the

24. Hubble Space Telescope image of the 'Pillars of Creation', a star-forming region in the Eagle Nebula in the Milky Way.

expansion of the Universe came about because he could detect the 'standard candles' of Cepheid variable stars in galaxies a few million light years away which allowed him to estimate their distances. However, his value for H_0 turned out to be too large. The 200-inch Hale telescope was later used to refine the measurement, reducing the uncertainties in H_0 to around 50 per cent. Hubble's high sensitivity and resolution, however, enabled it to detect Cepheid variable stars in galaxies hundreds of millions of light years away and, by extending the cosmic distance ladder to these greater distances, has reduced the uncertainties in determining the Hubble constant to within 10 percentage points.

In 2022, the HST discovered the most ancient and distant star so far seen (WHL0137-LS; or more poetically, *Earendel*, meaning 'morning star', named after J. R. R. Tolkien's mythical Middle-Earth star-carrying mariner). The light from Earendel was emitted only 900 million years after the Big Bang (7 per cent of the age of the Universe) and it was only detected by a chance alignment with an intervening cluster of galaxies which, through the action of gravitational lensing, has greatly magnified its light. The star has a high luminosity and is believed to have a mass of at least 50 solar masses.

One of the Hubble's most remarkable images is the Ultra Deep Field. It was made by pointing the telescope at a tiny, apparently empty area of sky, equivalent to the area covered by a sand grain held at arm's length (3 arcminutes across), within which the light was integrated for a total exposure of 11 hours. It is the deepest visible-light image we have of the Universe, a 2D projection of a slab of spacetime, stretching back 12 billion years, and collapsed into a single frame. Far from being empty, the tiny patch of sky contains around 10,000 faint galaxies. It showed that the youngest, most distant galaxies have different morphologies from those in the local regions of the Universe. These irregularly shaped galaxies appear to be merging and combining. The very youngest appear as dots a few pixels across, or smudges of coloured light, the embryonic nuclei of what would later become fully fledged galaxies. By counting the number of galaxies in the deep field and extrapolating it to the whole sky, it has been estimated that there are around 100 billion galaxies in the observable Universe. However, the problem with observing such distant stars and galaxies at visible wavelengths is that their light has been strongly redshifted to the infrared by the expansion of the Universe, and they start to disappear from view. In effect, the HST runs out of vision for these most distant objects.

The James Webb Space Telescope

Hubble's successor, the infrared James Webb Space Telescope (JWST; or simply 'Webb'), was launched from French Guiana on an Ariane 5 rocket on Christmas Day 2021. Squeezing the world's largest space telescope (6.5m aperture) inside the 5.4m-diameter nosecone of the launch vehicle proved to be a highly challenging and fortunately ultimately highly successful feat of engineering. Having been folded up like origami to fit in the nosecone, the telescope could only be finally assembled once in space. Unlike Hubble's low Earth orbit, which had made the telescope accessible to space shuttle repair missions, the remote and complex assembly of the Webb telescope had to proceed flawlessly. The most nerve-wracking part was unfurling the complex heat shield the size of a tennis court, emerging like the wings of a butterfly from a chrysalis. At Webb's heart is its gold-plated mirror, composed of 18 hexagonal gold-plated segments. At infrared wavelengths, a gold mirror is an efficient reflector. In this complex assembly process, there was no room for error; once in deep space there is no prospect of any repair missions. Webb is 1.6 million km from the Earth, four times the distance of the Moon, at the Sun–Earth *second Lagrangian point* (L2). L2 is a 'sweet spot' in space where the combined gravitational forces of the Earth and Sun are balanced by the centrifugal force of the spacecraft. There, Webb sweeps round the Sun, along the projected Sun–Earth line. Being in an L2 orbit is particularly advantageous for an infrared telescope operating at temperatures close to absolute zero because the environment there is much more stable than in a low Earth orbit where, moving periodically in and out of the Earth's shadow, it would be subject to large fluctuating temperature gradients.

The first Webb results, released in mid-2022, included a remarkable deep field image of the region around a cluster of galaxies (SMACS 0723; Figure 25). This image is a composite made at different wavelengths between 0.6 and 5 microns with

25. The James Webb space telescope's *First Deep Field* infrared image. It is a slice through spacetime, seeing galaxies as far back in time as only 300 million years after the Big Bang. Almost all the objects in the field are galaxies, including the faint smudges of light. In the centre of the image, the gravitational lens of an intervening galaxy cluster warps the light from the background galaxies into prominent arcs.

Webb's main imaging camera, NIRCam, with an exposure of 12.5 hours. It shows a tiny area of the sky, the infrared equivalent to that of Hubble's Ultra Deep Field. In the background there are extremely clear images of thousands of the faintest and most distant (and therefore oldest) galaxies that have ever been observed in the infrared. (The image contains a few nearby stars, the objects producing the sharp diffraction spikes.)

The light from the oldest galaxy seen here was emitted 13.5 billion years ago when the Universe was just 300 million years old. The frame of the image is centred on an intervening galaxy cluster 4.6 billion light years away which, acting as a gravitational lens, has magnified the images of the much more distant galaxies. The prominent arc-shaped distortions of the galaxy shapes in this image result from the powerful gravitational lensing effect of the cluster, similar to the familiar light bending and warping effect of a magnifying glass. The magnified images show details within the ancient galaxies such as individual star clusters. Webb is also equipped with a Mid-Infrared Instrument (MIRI) which is cooled below 7K using a cryocooler enabling it to operate in the 5–28-micron waveband. Images made at these longer wavelengths highlight the thermal radiation from hot dust in galaxies, which is associated with regions of star formation. By comparing Webb's observations in the two wavebands, it will be possible to study the evolution of star formation in galaxies at these very early times.

Other notable early JWST results include an infrared version of the 'Pillars of Creation' image (shown optically in Figure 24). At longer wavelengths the obscuring columns of dust and gas clouds become semi-transparent allowing Webb to see deeply into the heart of the nebula. This reveals myriads of hitherto unobserved young and still-forming stars. These young stars, no more than a few hundred thousand years old, periodically eject supersonic jets which blast into the surrounding gas to form bow shocks similar to the wake of a speedboat. Closer to home, Webb has made crisp images of the planets, including Jupiter showing its stormy surface, auroras, and faint ring, as well as the ice-giant planet Neptune, showing its surface features and remarkable multiple ring system.

Webb is the latest in a series of infrared space telescopes. The first mid- and far-infrared survey satellite was the InfraRed Astronomical Satellite (IRAS), dating from 1983. In a remarkable series of observations, IRAS discovered 350,000 infrared-emitting axies, many of them starburst types, and a hitherto unknown class of Ultra-Luminous Infrared Galaxies. Starburst galaxies undergo brief episodes of intense star formation, during which stars are formed at a rate much faster than is currently taking place in more quiescent galaxies like the Milky Way. Owing to the expansion of the Universe, the density of galaxies was once higher, and in the past galaxy–galaxy interactions and mergers were more frequent. Starburst galaxies are believed to arise from such galactic collisions, which can trigger the collapse of marginally stable interstellar clouds, initiating star formation. The strong stellar winds and UV light from the hot young stars blow dust out into space where it scatters and reradiates the light in the infrared. It is believed that the gravitational tidal interactions associated with galaxy mergers can drive fresh material deep down into the gravitational well of a supermassive black hole, triggering AGN behaviour.

In Webb's image of *Stefan's Quintet* (Figure 26), a group of interacting galaxies 300 million light years away, there is clear evidence of gravitational tidal interactions and mergers which are associated with starburst and AGN activity. The image shows galactic bridges and tails of stars and gas being pulled out from the galaxies by the tidal forces. Bright star clusters, starburst regions, and shock waves are visible in the luminous gas clouds lying between the central pair of galaxies on the right-hand side of the image. (The galaxy on the left-hand side is a chance foreground galaxy in the line of sight). The uppermost galaxy in the group harbours a supermassive black hole, an AGN accreting matter.

NASA's infrared Spitzer Space Telescope ('Spitzer') has been a mainstay of infrared observations such as observing the population of dust-shrouded starburst galaxies, from an epoch of the Universe called the cosmic *high-noon*. This was when the star formation rate in galaxies peaked, about 2–3 billion years after the Big Bang. Spitzer has also played a role in exoplanet research. In 2000, the Transiting Planets and Planetesimals Small Telescope

26. Stefan's Quintet. An interacting group of galaxies seen in infrared light from the Webb space telescope.

(TRAPPIST) in Chile discovered a pair of exoplanets orbiting a nearby red dwarf star, TRAPPIST-1, only 39 light years from the Sun. Spitzer was used to observe this cool star and found five more exoplanets. The sizes of the TRAPPIST planets have been measured by transit dimming to be Earth-like and one lies in the habitable zone of the star, where conditions are suitable for liquid water to exist on the planetary surface. By observing small variations in the timings of their transits, astronomers have been able to infer the masses of the TRAPPIST-1 planets which shows that these are rocky worlds. In another exoplanetary system, Spitzer was the first telescope to image directly the heat of a hot Jupiter, as well as discovering a previously unobserved ring around Saturn.

Mapping stars

On a clear moonless night, several thousand stars are visible to the naked eye. Astronomers have been making maps of the stars for centuries. Just as maps help us find our way around the Earth by pointing out useful landmarks, star charts help us find our way around the sky.

Mapping stars is part of the field of *astrometry*. Modern astrometry is concerned not only with the position and brightness of stars, but also with their kinematics and composition. The distribution of stars in space is described by six parameters: three space and three velocity coordinates. The stars we see from Earth are projected on the 2D surface of the celestial sphere. This gives us two spatial coordinates. To fix the third, distance, requires stereoscopic vision, or parallax. The smallest measurable parallax angles achievable with ground-based observations are around ten milliarcseconds but it is possible to improve this precision with space telescopes. The first dedicated astrometry satellite, Hipparcos, was launched in 1989. It used twin telescopes to measure parallaxes as small as one milliarcsecond which yielded accurate distances of 120,000 stars out to several hundred light years. The motions of stars projected on the celestial sphere are *proper motions*, the angular changes in a star's position with respect to the distant stellar background. Stellar proper motions therefore complement spectroscopic radial velocity measurements. When the proper motion, the radial velocity, *and* the distance to a star are all known, its position and 3D velocity in space can be found.

Gaia—surveying two billion stars

ESA's Gaia optical space telescope was launched in December 2013 and is currently making the most detailed multi-dimensional observations ever made of the position, distance, and movements of stars in the Milky Way. Like Hipparcos, Gaia's twin telescopes scan different parts of the sky, as the spacecraft slowly spins at its extremely dark Lagrangian L2 orbit, where it observes each star many times to pinpoint accurately its position and velocity. At Gaia's heart is a 1-billion-pixel camera, consisting of a mosaic array of CCDs generating an enormous amount of data. Transmitting all the raw data back to Earth would be prohibitively slow, so the important parameters such as stellar positions and magnitudes are extracted by on-board computers. The starlight is split into red and blue bands which allow the temperature, size, and chemical composition of individual stars to be inferred photometrically. The radial velocities of the brightest stars are measured with a high-resolution spectrometer. Even with data compression, the information transmitted back to Earth is equivalent to over ten DVDs worth per day.

Gaia's third data set, released in 2022, is the largest catalogue of astronomical data ever produced, and contains exquisite information on two billion stars, giving their temperatures, masses, ages, compositions, and types. These include many binary and variable stars. Gaia is also able to detect 'starquakes' in massive stars yielding data informing on their internal dynamics, temperatures, densities, internal rotations, and compositions. A star is essentially a hot ball of plasma, in which the inward pull of gravity is balanced by the outward pressure, a force balance that maintains the hydrostatic equilibrium of the star, keeping it inflated at a more-or-less constant size during its lifetime. Many stars, the Sun included, experience stellar oscillations about this equilibrium and, like the various oscillatory modes of a beaten drum-skin, they exhibit many possible vibrational modes. When a star's gas vibrates radially in and out, for example, it changes the luminosity of the star, making it 'blink' slightly. This motion has been observed in the Sun, where the almost imperceptibly small oscillations have informed on its internal structure. Just as seismology teaches us about the interior of the Earth, astroseismology informs us on the interior of stars. Different stars ıibit a great diversity of amplitudes of these vibrations which is ound to depend on their masses and evolutionary stage.

Gaia's data also enable 3D maps to be drawn of the positions and motions of vast numbers of stars, for example, the all-sky star map (Figure 27), showing the trails of stars in the solar neighbourhood projected 400,000 years into the future.

Gaia's data have also been used to project the positions of stars *backwards* in time. This has revealed the ancient history of our galaxy, and has shown evidence of past acts of galactic cannibalism. Galaxy–galaxy gravitational tidal interactions, such as that taking place in the Webb image (Figrue 26), produce gas and dust streamers bridging vast distances of intergalactic space. These and numerous other observations support the view that big galaxies get even bigger by eating up (or merging with) smaller ones. When a small galaxy (the victim) is consumed by a big galaxy (the perpetrator), the victim's stars will, over time, be assimilated and comingle with those of the perpetrator. Such acts of galactic cannibalism do not simply get obliterated from the history books; they leave traces of forensic evidence. For example,

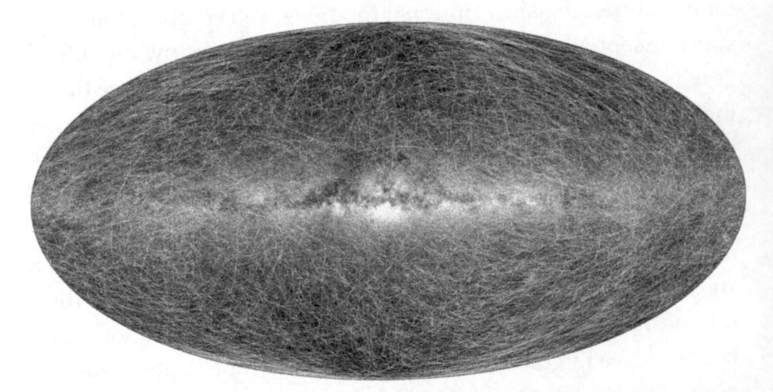

27. Gaia satellite's all-sky map of the trails of stars out to a distance of 300 light years, projected 400,000 years into the future. The plane of the Milky Way lies horizontally in the centre.

the victim's stars may be of a different age or composition to those of the perpetrator. So even if the stars become merged spatially in the final fattened-up galaxy, they can still be identified spectroscopically. If the victim merged, say, with a perpetrator on a counter-rotating orbit, a memory remains in the kinematic velocities of the stars. In Gaia's maps, astronomers have found evidence that around ten billion years ago, a dwarf galaxy known as Gaia-Enceladus punched its way into the Galactic disc, sending out gravitational ripples as the victim oscillated above and below it a few times before finally being pulled in by the overwhelmingly strong gravity of the Milky Way. Spectroscopic observations have revealed the existence of a special group of counter-rotating stars in the Milky Way's halo whose spectra have a different mix of chemical elements compared with their neighbours, evidence pointing to an ancient external galaxy collision.

Hunting for exoplanets

In 2009 NASA launched Kepler, a space mission designed to search for exoplanets. Kepler focused on exoplanets in the 'habitable zones' of their stars, where liquid water may exist, making them potentially suitable for the development of life. Kepler used the transit method, one that has turned out to be a productive way of discovering new worlds. To date, over 5,300 exoplanets have been discovered, of which over half were discovered by Kepler. Although the Kepler mission has now ended, computer algorithms and the Planet Hunters citizen science projects continue to sift through Kepler's data for evidence of planetary transits, filtering out many types of false positives. Once evidence for genuine exoplanets is found, these data are analysed to extract parameters such as the orbital and planetary radii, and the possible presence of multiple planets.

The first all-sky exoplanet transit telescope, NASA's Transiting Exoplanet Survey Satellite (TESS), was launched in 2018. It carries four wide-field telescopes, and is able to survey an area of sky 350 times larger than that viewed by Kepler. Its purpose is to identify a large sample of small planets which are suitable for follow-up observations; to date TESS has found over 6,000 candidate exoplanets, of which almost 300 have been confirmed as new at the time of writing. The criteria used in confirming a candidate exoplanet are stringent. These include observing multiple regular dips in a star's brightness, cross-checks with other telescopes, possibly at other wavelengths and with different techniques such as the radial velocity method, ruling out artefacts in the data or extraneous causes such as changes to a star's brightness caused by the movements of starspots or orbiting non-planetary bodies such as brown dwarfs, asteroids, or dust clouds.

The Webb telescope has a Near-Infrared Imager and Slitless Spectrograph (NIRISS) infrared spectrometer which was used in 2022 to measure the most detailed spectrum of the atmosphere of an exoplanet (WASP-96b) ever made. The planet was observed in transit across the face of a Sun-like star, some 1,150 light years from us, and is a hot Jupiter orbiting close to its star. Webb measured the spectrum of light transmitted through the exoplanet's atmosphere. The range of wavelengths studied (0.6–2.8 microns) brackets the spectral lines of water molecules, and the results showed clear evidence of water. The temperature on WASP-96b is in excess of 500°C, so the water must be in the form of steam. Webb also found evidence of haze and clouds on this planet.

Exoplanet research is an extremely active area, new results pour in from telescopes on the ground and in space. The thousands of exoplanets now confirmed show a highly diverse zoology. Masses range from less than 0.1 Earth masses for rocky worlds, to 10,000 Earth masses for hot Jupiters orbiting close to their parent stars, tidally locked and blasted by supersonic planet-wide winds, as well as cold gas giants orbiting further out. The orbital periods vary from as short as eight hours for hot molten lava planets, to over 100 years for the gas giants.

Ultraviolet observations

The wavelengths of ultraviolet light are longer than the spacing of atoms in solids, which means that UV can be reflected by finely polished mirrors. Hubble's mirror was polished to within 3nm for this purpose. Ultraviolet photons are emitted by objects with temperatures up to 100,000°, which include massive young stars, the accretion discs around gravitationally collapsed objects, AGN, and supernovae. Many chemical elements have atomic resonance lines in the UV, and observations of these inform on the chemistry of the interstellar medium.

In the 1960s NASA launched three Orbiting Astronomical Observatories (OAO) satellites. Of these, the Copernicus (OAO-3) observatory carried a high-resolution UV spectrometer and measured the spectra of many hot stars. The first space observatory to be operated remotely was the International Ultraviolet Explorer (IUE) of 1978 and has observed young hot stars, their stellar winds, hot white dwarf stars, and many active galactic nuclei. A large UV sky survey conducted in 2003 by NASA's Galaxy Evolution Explorer (GALEX) telescope, catalogued several hundred million sources including hot young stars in galaxies, which have improved our understanding of stellar evolution and nucleosynthesis.

The X-ray Universe

The first observations of cosmic X-rays were made from captured German ex-World War II V2 rockets and from balloons. Other suborbital rockets carried ionization chamber detectors above the atmosphere to observe the X-ray sky for a few minutes. An X-ray detection involves an incoming photon passing through a gas-filled chamber, ionizing the gas, and creating a wake of ions and electrons in the gas. This is recorded as an electrical pulse. Early observations of this type targeted the Sun, but our star was found to be only a weak source of X-rays. However, in 1962 an Aerobee rocket flight detected the first really bright X-ray source, Scorpius X-1 (Sco-X-1), an extraordinary object a billion times more luminous than the Sun in X-rays (or 10,000 times the overall luminosity of the Sun). This observation was highly influential in starting X-ray astronomy, a field that has now shown us that the Universe is teeming with extraordinary X-ray-emitting objects such as X-ray binary (XRB) sources containing compact objects such as black holes, neutron stars, supernova remnants, galaxies, quasars, and the vast atmospheres of hot gas bound in clusters of galaxies. Very hot objects with temperatures from a million to a billion degrees radiate X-ray photons with energies between 100 eV and 100 kilo electron volt (keV).

In 1970 the first X-ray survey satellite, Uhuru (meaning 'freedom' in Swahili), was launched from a site in Kenya. Uhuru's catalogue contains a large number of XRBs, the brightest type of X-ray sources in the Milky Way. An XRB consists of a compact object (white dwarf, neutron star, or black hole) and a normal star orbiting around their common centre of mass (Figure 28). A star in a binary system is bounded by a teardrop-shaped surface called the *Roche lobe*, shown distorting the envelope of the star to the right of Figure 28. The Roche lobe defines the region within which the stellar gas is gravitationally bound to the star. If the gas extends outside of the star's Roche lobe, it can be captured by the compact object. In an XRB, matter is stripped from the normal star (the donor) and transferred to the compact object via the intermediary of a hot X-ray-emitting accretion disc. The disc radiates an enormous amount of energy (several tens of percent of the rest mass energy of the matter). When the XRB also has jets and is a radio source, it is known as a a *microquasar*, essentially a miniature version of an AGN. One of the brightest XRBs in the sky is Cygnus X-1 (Cyg X-1), the first black hole ever identified, which is believed to weigh around 15 solar masses. Cyg X-1 is orbiting a bright blue supergiant star (HDE226868) with a

28. An X-ray binary, showing the transfer of matter to a compact object from a companion star.

5.6-day period. The X-ray intensity is observed to vary rapidly on timescales of about a millisecond, implying that its emitting region is smaller than 300km across.

One of Uhuru's most important discoveries was of the X-ray emission from hot gas in clusters of galaxies. Clusters of galaxies are some of the largest gravitationally bound entities in the Universe. The Coma cluster, for example, has a diameter of 20 million light years and contains roughly 1,000, mostly elliptical, galaxies. This cluster had been studied in 1937 by Fritz Zwicky. His spectroscopic measurements showed that the galaxies are moving around too quickly for the Coma cluster to be stable if the only mass present was the visible mass in galaxies. Zwicky concluded that there must be a large amount of extra, unseen mass present to provide enough gravity to prevent the galaxies from flying apart, and he named this missing mass 'dunkle materie' (or dark matter). However, dark matter must be of a ındamentally different nature to the baryonic mass in protons and neutrons (baryons) of so-called normal matter, because it does not emit, absorb, or reflect light.

Zwicky's dynamical argument was based on his spectroscopic observations of the high velocities of galaxies in the cluster. However, a different method of determining the total gravitational mass in clusters of galaxies comes from X-ray observations. Large clusters of galaxies are permeated by a million-degree-hot intracluster medium consisting mainly of ionized hydrogen and helium. High temperature plasmas emit X-rays as *bremsstrahlung* (braking) radiation which is produced by the sudden deflections of fast-moving plasma electrons encountering the strong electric fields of the plasma ions. The gas in a cluster of galaxies is in a state of hydrostatic equilibrium resulting from a force balance: gravity pulling the gas inwards, and thermal pressure pushing it outwards, as it is in a star. So by balancing the forces, the gravitational potential of the cluster and its total gravitational mass can be inferred from a knowledge of the gas density and pressure, both of which come from X-ray observations. The results showed that large clusters of galaxies weigh roughly ten times more than can be accounted for by the amount of visible matter they contain. In other words, around 90 per cent of the cluster mass is not normal baryonic matter, but dark matter.

X-ray telescopes

The first X-ray images were crude and made at low resolution using collimators placed in front of ionization detectors. The first focusing X-ray telescope (XRT) was a grazing-incidence design by German physicist Hans Wolter (Figure 29(a)). While energetic X-ray photons penetrate different materials to varying degrees, they can bounce off ultra-smooth metal surfaces at grazing incidence, like stones skipped across a smooth lake. For this, the surface has to be atomically smooth since any bumps, even a few atoms high, will scatter X-rays and spoil the image. The aperture

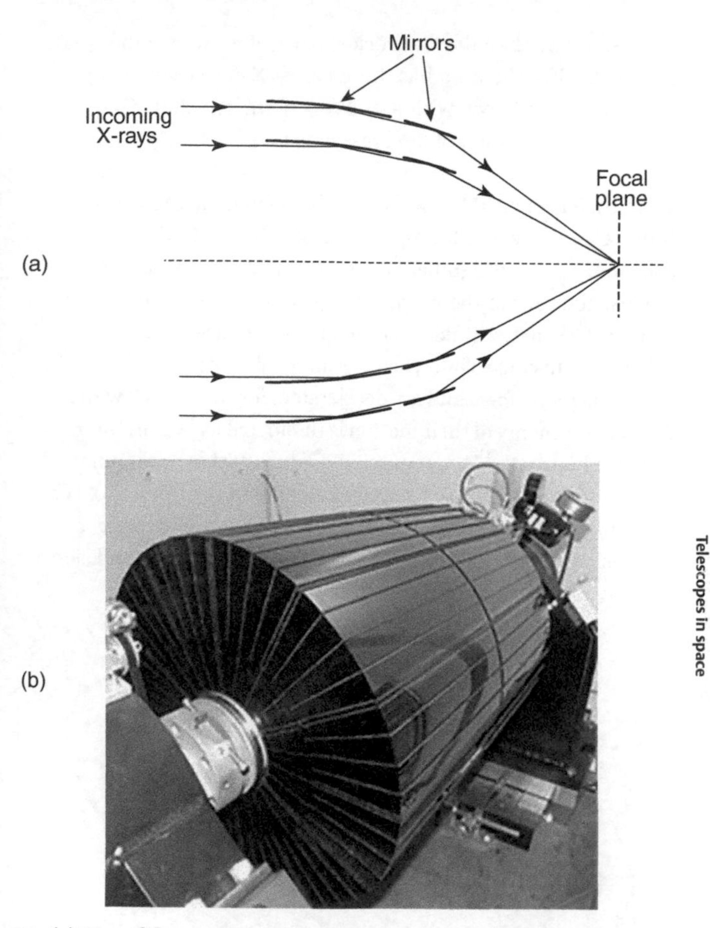

29. (a) Two of the many nested grazing incidence mirrors in a Wolter X-ray telescope; incoming X-rays bounce off the inclined mirror surfaces to the focus. (b) The Chandra X-ray telescope.

of an XRT is formed from multiple sets of cylindrical Wolter mirrors, nested like Russian dolls. In 1999, NASA launched the Chandra X-ray Observatory ('Chandra', named after Subrahmanyan Chandrasekhar), a telescope ten billion times more sensitive than the first rocket-borne detectors of the 1960s (Figure 29(b)). Chandra has the sharpest X-ray vision of any telescope so far flown, with its 1.23m aperture Wolter XRT achieving a resolution of 0.5 arcseconds.

Chandra has imaged the X-ray emission from the remarkable Bullet Cluster, two colliding clusters of galaxies. In this titanic collision, there are significant observational differences between the distributions of the intracluster gas (from X-ray observations), and the dark matter (inferred from gravitational lensing). When two clusters collide, their gas atmospheres interact via the electromagnetic force and are decelerated, forming shock waves. The kinetic energy of their motion is dissipated as heat in the gas which enhances the X-ray emission. Figure 30 shows Chandra's observations of the intracluster gas component (in greyscale). Where the two gas components of each cluster have interpenetrated, a bullet-shaped bow shock wave has formed on the right-hand side of the image, reminiscent of the wake of a speedboat on water.

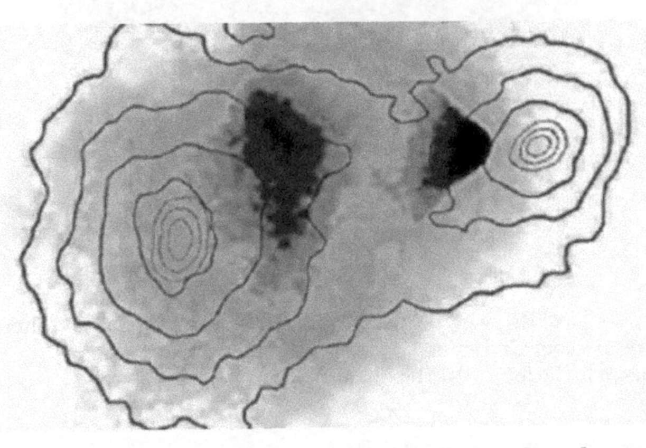

30. The Bullet Cluster. Overlay image of the collision of two clusters of galaxies. Chandra's X-ray observations are shown in greyscale with the two darker shaded peaks indicating stronger emission from the gas. The X-ray peaks are displaced away from the two peaks in the total gravitational mass (contours).

A contour map of the total gravitational mass distribution of the colliding clusters (inferred from gravitational lensing) is superimposed over the X-ray map. The two mass density peaks indicate the positions of the dark matter clumps. There is a spatial mismatch between the distributions of dark matter and the gas. In the collision of these two vast clusters of galaxies, the gas atmospheres appear to have been decelerated and are now lagging behind the clumps of dark matter which seem to have passed through themselves collisionlessly. The bullet cluster observations provide evidence that dark matter does not appreciably interact either with itself or with normal matter, except via gravity.

Gamma-ray astronomy

Gamma rays cover an enormous energy range, from 100keV to 10^{20}eV (ten million times higher in energy than has been achieved in the world's most powerful particle accelerator, the Large Hadron Collider (LHC) at the European Organization for Nuclear Research (or CERN)). Such energetic photons cannot be focused by the familiar reflecting mirrors or refracting lenses in a normal telescope—they simply pass through solid materials as if they are bullets fired through tissue paper. Instead, astronomers study gamma-ray interactions with matter in detectors that would not be out of place in a high-energy physics laboratory.

Three ways that high-energy photons interact with matter are through the photoelectric effect, *Compton scattering*, and *pair production*. The photoelectric effect is used to detect lower energy photons, as in a CCD detector. For medium energies, up to 30 million electron volt (MeV), Compton scattering predominates. The American physicist Arthur Compton discovered that high-energy photons behave like billiard balls so that when they collide with electrons energy is shared out between them. This property is used to detect gamma rays and forms the basis of Compton gamma-ray scintillators, materials which absorb the energy of ionizing radiation and convert it into visible photons.

In a two-stage scintillator, a photon is first scattered to lower energies and then absorbed in a second scintillator to produce a photoelectron. At the highest photon energies, interactions with matter are dominated by pair-production events. When a gamma ray grazes past an atomic nucleus, the nuclear electric field causes it to transform into a pair of particles: an electron and a positron. A positron is an antimatter particle, a positively charged electron. The pair travels on through the detector, leaving a tell-tale trail of tracks. For this process to occur, the photon energy must exceed 1.02MeV, twice the rest mass energy of the electron. The electron and positron subsequently annihilate to produce two gamma rays which can either Compton scatter or be absorbed photoelectrically.

NASA's Compton Gamma-Ray Observatory (CGRO), active between 1991 and 2000, was the first large space observatory to carry a variety of X- and gamma-ray detectors including a gamma-ray spectrometer to measure the nuclear spectral lines up to 10MeV produced by radioactive decay and so identify chemical nuclei. CGRO also carried a pair-production gamma-ray telescope, a *spark chamber*. This is a gas-filled chamber containing an array of electrically charged metal plates which show up the tracks of pair-production particles. The direction of an incoming photon is deduced from back-projecting the tracks.

Other gamma-ray sources include the acceleration of cosmic rays ploughing into the interstellar medium of a galaxy, or sites of the acceleration of particles in the shock fronts around supernovae explosions. Gammas are also produced in the fleeting billion-degree temperatures inside supernovae explosions, by matter–antimatter annihilations, and by the process of *inverse Compton scattering*. In this, low-energy photons receive a large energy boost when they collide with very hot electrons, converting them into gamma rays. Such conditions are expected in the relativistic jets of AGN.

The most detailed images of the gamma-ray Universe have come from NASA's Fermi Space Telescope (Fermi) of 2008, named after the Italian nuclear physicist Enrico Fermi. Fermi can image gamma rays with energies up to 300 giga electron volt (GeV) with an angular resolution of 0.1°, using its 20° field-of-view pair-conversion Large Area Telescope. Fermi's gamma-ray survey has resulted in a catalogue of some 5,000 objects identified as blazars, pulsars, and supernova remnants. There is a band of gamma-ray emission from the plane of the Milky Way, and a particularly bright source in the galactic centre near the supermassive black hole, Sgr A*. A significant number of Fermi's sources are so far unidentified. One of the most striking discoveries was of two vast gamma-ray-emitting gas bubbles, located symmetrically on either side of the centre of the Milky Way, extending tens of thousands of light years from the galactic plane. These so-called 'Fermi bubbles' extend even further from the galactic plane than the MeerKAT radio-emitting filaments (Figure 23), and are thought to result from past explosions from the region near the central supermassive black hole.

Chapter 5
The dynamic Universe

Gazing up at the star-studded sky on a clear dark night, it is easy to think of the heavens as being static. The only barely discernible motion is the gentle wheeling of the constellations on the celestial sphere, against which the Moon and planets slowly weave their wandering courses. The stars are changing, but imperceptibly so on human timescales. Stars and galaxies evolve on timescales measured in millions or billions of years. Occasionally a star might suddenly flare up as a nova (visible at a rate of about one per decade) or explode as a supernova (seen about once a century). Once thought to be rare events, modern *time-domain astronomy* has shown us the opposite. Every second a supernova explodes somewhere in the Universe. Things can and do change rapidly, often quite suddenly and without warning.

Astronomers search the sky for objects that move or suddenly change brightness. The objects that flare up on timescales of milliseconds are the supernovae, variable X-ray sources, magnetars, gamma-ray bursts, fast radio bursts, active galactic nuclei, and tidal disruption events. Rapidly varying objects are supermassive black holes and the stellar corpses of white dwarfs, neutron stars, and black holes. However, the problem with observing any transient event lies in catching it. For example, during a meteor shower, a 'shooting star' can appear at any moment, anywhere in the sky. To catch sight of one, a good strategy is simply to rely on the wide field of the naked eye. For the same reason, wide-field telescopes are the most effective for this type of observation.

A global system dedicated to catching transients is NASA's Catalina Real Time Transient Survey, an observing network based on three telescopes with apertures no bigger than 1.5m, located in the USA and Australia. Catalina has discovered thousands of optical transients, mostly AGN with supermassive black holes, supernovae, cataclysmic variables (accreting binary stars that emit sporadic outbursts), as well as many asteroids and Near-Earth Objects (NEO). (An NEO is defined as an object coming within 1.3AU of the Sun.) Another dedicated telescope is the Zwicky Transient Facility (ZTF) at Palomar Observatory. The ZTF telescope is the venerable and now robotic Samuel Oschin Schmidt camera, which was refurbished in 2017 with a cooled CCD camera, 100 times more sensitive than the original photographic plates it once used. ZTF can image 47 square degrees of sky at a time, equivalent to around 250 full Moons, and can survey the entire northern sky in two nights. The telescope is currently detecting over 10,000 explosive transients a year and creating a torrent of data including hundreds of thousands of nightly alerts. Most of these are false and, to filter them out and find the 'needles in the haystack', the rapidly growing field of *machine learning* is being developed to classify the genuine events as supernovae, asteroids, and objects that can be followed up spectroscopically.

Transients are discovered by scanning an area of the sky, storing the digital image, moving to another field, and then returning to repeat the first scan. Any changes are revealed by subtracting successive images of the same region. When a 'target of interest' is identified, its coordinates are sent to telescope networks, for further observation. The most important information about a transient comes during the earliest stages, and it is vital to capture data from such an object as quickly as possible, before it fades from view.

Supernovae

On 23 February 1987, a supernova, SN1987A, exploded in our nearby galaxy companion, the Large Magellanic Cloud. It was the brightest since Kepler's supernova of 1604, and it continues to be important because it has been studied in great detail for three decades using the whole panoply of modern telescopes and instruments, studies that have informed us on the final stages of stellar evolution. SN1987A was a Type II supernova which results from the gravitational collapse of the core of a massive star weighing more than 8 solar masses. The explosion was near enough for astronomers to identify that the progenitor had been a 20-solar-mass blue supergiant star which, afterwards, disappeared. The light curve of the expanding shell of debris took 80 days to reach maximum brightness, and then remained at roughly constant luminosity for around two months. A hundred days after the explosion, spectroscopic observations showed that the light was dominated by the radioactive decay of nickel 56 to cobalt 56 to iron 56, a decay chain that gives off gamma rays. Being released in an expanding optically thick shell, the gamma rays were scattered in the material, reradiating the energy in the optical and infrared wavebands. The gammas, however, could not be observed until the shell had expanded and become transparent enough to allow them to escape. The supernova remnant stayed visible in UV, optical, and infrared wavebands for around three years.

A remarkable late image of SN1987A combines data at three wavelengths (Figure 31). It shows a sharply defined luminous ring of excited gas centred on the explosion, produced by the UV flash exciting atoms in the surrounding gas. Three overlapping rings of material have been observed that are believed to have been blown off during the star's final years. The slower moving shock wave from the explosion eventually caught up with the luminous ring which, when they collided, heated the gas and enhanced the X-ray

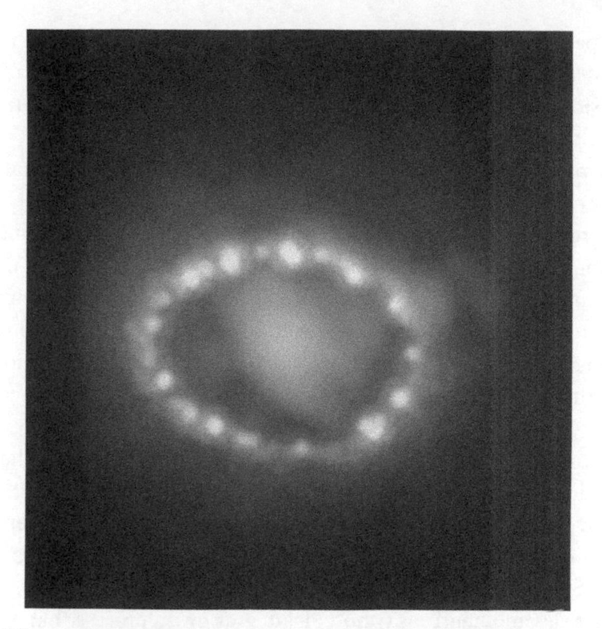

31. Three superimposed images of supernova 1987A showing the visible light image (from the HST) with the spectacular 'ring of pearls' produced by the blast wave striking clouds of plasma surrounding the explosion, the central diffuse millimetre wave emission seen by the ALMA telescope, and the smooth doughnut-shaped ring of X-ray emission from Chandra.

luminosity. The remnant also emitted radio synchrotron emission, from electrons accelerated in the shock wave. Is there an associated neutron star? None was identified at first but, after many years of searching, astronomers using the ALMA telescope in 2019 believe that they have observed it at submillimetre wavelengths. The neutron star is thought to have been lying hitherto unseen, shrouded in dust.

These observations have enabled us to frame the main events leading to a core-collapse supernova. Towards the end of their lives, the outer layers of massive stars expand and they become red supergiants. The high core temperatures in these stars take them past the carbon and oxygen burning stage, initiating a series of fusion reactions synthesizing higher mass elements in the periodic table such as neon and silicon in onion-skin-like shells surrounding an iron core. Fusion reactions are *exothermic*, namely they release energy. However, as the massive star produces increasingly heavy elements, it encounters a law of diminishing returns. The energy yield of fusion reactions drops, eventually becoming zero at iron. Thus, to make elements heavier than iron requires *endothermic* nuclear reactions which require a net energy input. Therefore, once the iron core has formed, fusion reactions in the centre cease.

At this point, events proceed rapidly. The iron core exceeds the critical Chandrasekhar mass and begins to collapse, releasing gravitational energy which causes the temperature to shoot up several billion degrees. The radiation trapped in the core is so energetic that the photons break up iron nuclei into alpha particles and neutrons (photodisintegration), a process that absorbs so much energy the central pressure falls precipitously, turning the contraction into a free-fall. As temperatures skyrocket, the helium nuclei split into protons and neutrons. The protons convert into neutrons via the weak nuclear interaction, releasing an enormous flux of neutrinos in forming the neutron star. The sudden onset of the neutron degeneracy pressure stabilizes the core, stopping the collapse dead in its tracks. Meanwhile, unaware of the now-stable core, the outer envelope of the star continues its headlong infall. On striking the core, the envelope rebounds violently, flinging a rich cocktail of elements into the interstellar medium at speeds of 10,000km per second and generating powerful shock waves in the gas. This is a nuclear explosion that is an octillion (10^{27}) times more powerful than a hydrogen bomb.

In a Type Ia supernova, the progenitor is a carbon–oxygen white dwarf star accreting matter from its companion star in a close binary system, which takes it over the Chandrasekhar stability limit. (Types Ib and Ic are different still and are believed to be variants of core-collapse types, where much of the stars' outer envelopes have been stripped away.) When the matter from a companion star accretes onto the surface of the white dwarf, it is heated strongly, and undergoes a flaring thermonuclear burn, brightening up sporadically as a *nova*. In this way, the white dwarf's mass will steadily build up, increasing until it reaches the critical Chandrasekhar mass. At that point, the temperature rises, carbon is ignited, and the star collapses rapidly, triggering runaway thermonuclear reactions. In a few seconds the entire star unbinds and explodes as a supernova. Because the instability threshold occurs at a fixed mass, Type Ia supernovae reach consistent peak luminosities (about five billion times the luminosity of the Sun). This renders them visible over great cosmic distances, where they are used as *the* standard candles that underpin modern cosmology and enable us to assess the scale of the Universe.

Gamma-ray and X-ray transients

In the 1960s, the USA signed a test ban treaty with the USSR to eliminate atmospheric nuclear weapons testing. A nuclear air burst produces a large flux of gamma rays and, to monitor Soviet compliance with the treaty, the USA launched a number of Vela satellites carrying gamma-ray sensors. By timing the arrival of a gamma-ray pulse at different satellites, the position of the source can be pinpointed. The satellites did indeed detect bursts of intense gamma rays, however these came not from the Earth, but from the cosmos. The Vela results were declassified and published in 1973.

Gamma-ray bursts (GRBs) are the brightest events in the gamma-ray sky. They can appear at random anywhere in the sky about once a day, and can last anything between ten milliseconds and twenty minutes. The Compton survey showed GRBs to be uniformly distributed across the sky and therefore not especially associated with the Milky Way. This meant that they could have come from either weak local sources or strong ones much farther away. There are two types: short bursters and long bursters, with a break point at around two seconds. The long bursters are thought to come from core-collapse supernovae forming either neutron stars or *hypernovae* (also known as collapsars, where the core collapses directly to a black hole). In 1997, four decades after GRBs were first observed, the fading afterglow of a long-burst event, GRB970508, was at last identified optically with a faint star-forming galaxy, allowing its distance to be measured. This was six billion light years, indicating that this and other GRBs release massive amounts of energy. On 9 October 2022, the 'Brightest Of All Time' (BOAT; GRB221009A) was observed using both ground and space telescopes. The gamma-ray flash was so intense that it temporarily blinded several space detectors. BOAT was a long-burst GRB, associated with the explosion of a rare type of massive rapidly rotating star. Observations pinpointed the explosion to a galaxy about two billion light years away. A star in this galaxy is thought to have collapsed directly into a black hole, whilst simultaneously ejecting material into space as two opposite and narrow relativistic jets, one of which happened to point towards us. The flux of energetic photons arriving from the GRB was large enough to disturb perceptibly the Earth's upper atmosphere, ionizing the gas, and modifying its radio propagation properties. This impact also produced very-low-frequency oscillating currents in the Earth's crust, coupled via the Earth–ionosphere waveguide.

The short-burst GRBs are believed to have a different origin; there is evidence that these are produced by neutron star to neutron star mergers. This idea gained credence in 2017 when the LIGO-Virgo consortium observed the first multi-messenger gravitational wave/electromagnetic wave merger (GW170817) which coincided with a short GRB seen by the Fermi gamma-ray satellite. This event was also seen by NASA's Neil Gehrels Swift Observatory. Swift, launched in 2004, carries a Burst Alert Telescope to image GRBs, and observes about 100 a year with a *coded aperture telescope.* This type of telescope is a sensor array fronted by a lead mask which admits gamma rays through pinholes placed in a fixed random pattern. The recorded images are later analysed to reconstruct an image which locates the source. Swift also carries Wolter X-ray and UV/optical telescopes which can be slewed rapidly to observe the GRB afterglow, and can broadcast transient alerts to other telescopes. Swift has made many exciting discoveries, including that of the most distant GRB seen so far, 13 billion light years away.

X-ray transients are associated with the transfer of mass from a star to a compact object in an X-ray binary (XRB), as shown in Figure 28. There are two varieties of XRBs: low- and high-mass types such as Cyg-X-1. Low-mass X-ray binary systems (LMXBs) have lightweight stellar companions, such as the half-solar-mass donor star which is the companion to the accreting neutron star and the brightest X-ray source in the sky, Sco-X-1 (its optical counterpart is a bright highly variable star, known as V818 Sco, but most of this light comes from the accretion disc around the neutron star). LMXBs do not pulsate regularly but produce a quasi-steady X-ray emission, varying on timescales of minutes to months, punctuated by sporadic and dramatic X-ray outbursts (so-called 'X-ray bursts') that rise within seconds and decay within minutes. This behaviour is explained by the accretion of hydrogen and helium onto the surface of the neutron star releasing energy and heating the gas to high temperatures. The hydrogen readily burns on the surface of the neutron star, increasing the X-ray emission, but a critical amount of helium has to accumulate before a thermonuclear 'flash' is triggered, which we observe as an X-ray burst. Pulsations are also seen. Matter falling onto a magnetized neutron star is funnelled towards the magnetic poles where it releases energy in X-ray emitting hot spots. As the star spins, these spots move in and out of the observer's line of sight and thus appear as regular pulsations in the X-ray signal.

Tidal disruption events

Isaac Newton's universal law of gravitation states that the attractive gravitational force between two masses is proportional to the product of the masses and inversely proportional to the square of their separation. Newton used the law to explain the ocean tides, caused by the gravitational influence of the Moon. In the open seas there are two tides a day. On the side of the Earth facing the Moon the water is nearest to the Moon and pulled towards it most. Being able to move, the water bulges out to make a tide. On the opposite side, being farther from the Moon, the water is pulled less strongly and so is 'left behind' by the Earth, bulging out to make the second tide. These relatively weak tidal forces move the water by no more than a few metres. But when the orbit of a star brings it close to a supermassive black hole, tidal forces can rip the whole star apart.

If a star wanders too close to a black hole, tidal forces deform the star's facing surface strongly, causing it to bulge out towards the black hole, and elongating it into a cigar shape. Once the stellar material protrudes outside the Roche lobe, it can be captured by the black hole. The material then further elongates, eventually forming a filament winding around the hole. The filament fragments into separate pieces which fall into the hole producing a string of observable transient radio, optical, X-ray, and gamma-ray flares. This star-shredding process is a *tidal disruption event*, and many are seen in surveys. A classic nearby and small-scale example of this fragmentation process took place in 1992 when Comet Shoemaker-Levy 9 passed so close to Jupiter that the planet's tidal forces tore it apart, forming a line of cometary pieces, all of which subsequently fell into the giant planet.

Magnetars

One of the most mysterious and terrifying objects in the Universe is a rare type of pulsar called a *magnetar* (magnetic star). Magnetars are the most highly magnetized objects known in the cosmos and generate powerful outbursts of X-rays and gamma rays. The details of how magnetars form are not understood but, like pulsars, they are thought to be produced in the collapsing cores of massive stars in supernovae. The enormous magnetic fields are likely to arise from a magnetic-line winding-up process, in which accretion from a companion is combined with rotation. A magnetar has a magnetic field strength of a quadrillion (10^{15}) gauss. To put this in perspective, the Earth's field is about 0.5 gauss, a common refrigerator magnet has around 100 gauss, in a pulsar the field strength is 1 trillion gauss, but in a magnetar it is 1,000 times stronger still. Near to a magnetar the field is so strong that it can distort matter by tearing molecules apart and stretching atoms into rod-like shapes called *pencils*. Even if you were 1,000km from a magnetar, its magnetic field would distort the atoms in your body.

Magnetars generate some of the strongest known electromagnetic transients. As is evidenced by the observed glitches in pulsar periods, the stresses in the neutron crust can build up, slip, and fracture, generating extremely powerful *starquakes*. Starquakes are analogous to the build up and release of stresses in Earth's tectonic plates, which result in earthquakes. Magnetar starquakes are associated with powerful emissions of X-rays and gamma rays. One of the most powerful starquakes was detected from magnetar SGR1806-20, an object only about 12km across and 50,000 light years away, on the far side of the Milky Way. In a burst lasting only a tenth of a second, this starquake released more energy than the Sun does in 100,000 years. In 2013, a magnetar (PSR 1745-2900) was discovered orbiting less than a light year from the supermassive black hole in the centre of the Milky Way, Sgr A*, producing transients at X-ray, gamma-ray, and radio wavelengths.

Fast radio bursts

The 64m radio telescope dish at Parkes Observatory in New South Wales, Australia, started operating in 1961 and over the years has discovered a record number of pulsars. It also picked up the first of a mysterious type of brief radio transient, a *fast radio burst* (FRB). In 2007, radio astronomers Duncan Lorimer and David Narkevic had been reanalysing some six-year-old pulsar data when they came across a strange signal—an extremely bright millisecond-long single radio burst. The resolving power of the Parkes dish was not great enough to determine the position of what became known as the 'Lorimer burst' well enough for optical identification, but the signal had an incriminating fingerprint—an unusually large *dispersion measure.*

Two key observables of a pulsar are its pulse period and dispersion measure. Pulsars emit sharp pulses which, on their journey through the interstellar medium, are smeared out in time so that the low-frequency waves in the pulse lag behind the higher frequency ones and arrive later. Pulse dispersion is caused by the interaction of the wave with electrons in the interstellar medium, slowing it down. The observed pulse broadening, the dispersion measure, relates directly to the number of electrons the ray has encountered along its path. With a knowledge of the electron density in the galaxy, pulsar dispersion measures can therefore be used to infer the distances of pulsars.

What was unusual about the Lorimer burst was that its dispersion measure was much larger than that possible for any Milky Way pulsar. There is no line of sight through the interstellar medium with enough electrons that could explain it. Beyond the Milky Way's outer reaches lies a tenuous intergalactic medium, where the electron density falls away precipitously to a much lower level.

To account for the large dispersion, the Lorimer burst must therefore have been very distant and very powerful. In 2015, Parkes detected another FRB (FRB150418) which this time was also picked up by the Australia Telescope Compact (ATCA) radio telescope. ATCA pinpointed the position sufficiently precisely to associate the fading afterglow with a galaxy about six billion light years away. The pulse was reminiscent of a short GRB which is now believed to be produced by the merger of two neutron stars. If an FRB is itself associated with a GRB, this would have implied that an FRB is a one-off cataclysmic event. However, the Arecibo telescope discovered a different type of FRB, a *repeater* (FRB121102) in a dwarf galaxy 2.5 billion light years away. FRB121102 appears to go to sleep and wake up on a regular 156-day cycle, and was observed again in 2020 by the Chinese FAST radio dish. These observations imply that the energy in a single pulse is huge: a one millisecond FRB contains as much energy as that emitted by the Sun in a century.

The Canadian Hydrogen Intensity Mapping Experiment (CHIME) radio telescope saw its first light in 2017. It operates at low frequencies and is located in a radio-quiet zone in British Columbia's Rocky Mountains. CHIME is configured as a transit telescope consisting of four 100m-long near-half-cylindrical reflectors placed side by side pointing to the zenith. Over each of these reflectors are suspended strings of dipole aerials. Like LOFAR, CHIME is a radio telescope with no moving parts, and uses digital beam-forming technology to direct the beam to different parts of the sky. With its wide field of view and large aperture, CHIME is effective at discovering FRBs, which appear randomly in the sky. In its first year of operation, CHIME detected 535 new FRBs, most of these being the single-pulse variety in distant galaxies. These brief and mysterious radio beacons are not understood and are currently an active area of research. Some FRBs emit polarized radiation, indicating that they come from regions containing strong magnetic fields. Suggestions include the possibility that the massive energy releases they produce are associated with particle acceleration in the intense magnetic fields of magnetars or in compact object mergers involving black holes or neutron stars.

Near-Earth objects

Near Earth Objects (NEOs) include comets and asteroids as well as the so-called 'killer' objects that could potentially collide with the Earth. Fortunately, these objects with sizes of over 1km are extremely rare, yet our planet is continually being struck by much smaller objects the size of sand grains that burn up in the atmosphere in meteor showers and that bring us about 100 tonnes of matter every day. A potentially hazardous NEO is one larger than 140m that intercepts the Earth's orbit. An observatory dedicated to searching for these objects is the 2008 Panoramic Survey Telescope and Rapid Response System (Pan-STARRS), which operates a pair of wide-field 1.8m Ritchey-Chrétien survey telescopes in Hawaii. Pan-STARRS has detected many new solar system objects as well as variable stars, and supernovae. In 2017 it discovered Oumuamua, an object between 100m and 1km across and the first interstellar object ever seen passing through the solar system on an unbound hyperbolic trajectory. It has been speculated that Oumuamua may have been ejected from an exoplanetary system.

On 26 September 2022, NASA deliberately crashed a spacecraft, the Double Asteroid Redirection Test (DART), into a 163m-wide asteroid moonlet (Dimorphos) which is orbiting a larger asteroid, Didymos. This was an experiment to find out if the impulse from the collision could deflect the moonlet into a different orbit, and it was the first ever test of a planetary defence system that could potentially save us from the fate that befell the non-avian dinosaurs 66 million years ago, when three-quarters of Earth's species became extinct. The extinction event is generally believed to have resulted from the impact of a 10km-wide asteroid in the Gulf of Mexico, known as the Chicxulub impactor. Follow-up observations were made using ground and space observatories including both the Hubble and Webb telescopes. These showed an initial cloud of material ejected from DART's impact on Dimorphos and has confirmed that the collision has changed the moonlet asteroid's motion in space, reducing its orbital period by around 5 per cent.

The night sky, far from being quiescent, is punctuated by the movement or change in brightness of objects, sometimes across all wavebands. However, not all of these time-varying objects emit electromagnetic radiation and, in the next chapter, the theme will continue as we look at two new cosmic messengers.

Chapter 6
Multi-messenger astronomy

Neutrino astronomy

Neutrinos (or 'little neutral ones') are tiny, almost massless, elementary particles that were postulated by Pauli in 1930, named by Enrico Fermi in 1934, but not detected until 1956, when they were finally observed in a nuclear reactor by US physicists Frederick Reines and Clyde Cowan. The long delay between hypothesis and observation arose because the interaction between neutrinos and matter is so puny that a neutrino can pass through a light-year thickness of lead before interacting with an atom. So weak is the neutrino–matter interaction in fact that we are blissfully unaware of the vast numbers of neutrinos that pass unfelt through our bodies every second, day and night. Cosmic neutrinos are produced in the stars, supernovae, AGN, gamma-ray bursts, cosmic ray interactions in the atmosphere, and in the Big Bang, and they exist over a vast range of energies and fluxes. Although their feeble interactions make them hard to detect, when they *are* detected it means that many will have come from electromagnetically inaccessible regions of the Universe, making them highly informative messengers.

The first ever neutrinos detected from outside the solar system were from the supernova SN1987A in the Large Magellanic Cloud, an observation that really started high-energy neutrino astronomy. Bursts of neutrinos were detected several hours before the arrival of the first photons; arriving from the direction of southern skies, the neutrinos had to traverse the body of Earth before entering the detectors of the Kamiokande neutrino experiment in Japan and the Irvine-Michigan-Brookhaven (IMB) experiment in the USA, both located in the northern hemisphere. Each experiment saw a near-simultaneous, twelve-second-long burst of neutrinos. What caused the delay between the observed neutrino and photon signals? When the core of a massive star collapses a massive amount of gravitational energy is released, which is mostly carried away by neutrinos. These stream out from the collapsing core at almost the speed of light, and do not interact significantly with the rest of the star. The photons, on the other hand, interact strongly with the expanding cloud of ejecta from the explosion, and scatter multiple times before finally being released into space, thus accounting for the time delay.

Neutrinos exist in three 'flavours': electron, muon, and tau neutrinos, which interact with their respective paired elementary particles. A neutrino travelling through space morphs between the three flavours, a property known as *neutrino oscillation*. A neutrino emitted with, say, an electron flavour can, when observed later, be seen as a tau or a muon neutrino. An important source of astrophysical neutrinos is the series of fusion reactions which converts hydrogen into helium and powers the stars. When four protons come together to fuse into helium in a star, two of them must convert into neutrons via the nuclear beta decay process. This interaction also releases positrons and electron neutrinos. Each second, the Sun produces over 10^{39} neutrinos in its core which pass through the stellar surface in about a second, before flying off into space. The solar neutrino flux is so large that even on Earth, 150 million km away, 100 billion neutrinos zip through your thumbnail every second. From 1964 onwards, US astrophysicist John Bahcall developed the standard solar model, which predicts the solar neutrino flux on the basis of the conditions and known fusion reactions in stars. Bahcall worked with US physical chemist Raymond Davis, who set up an experiment to search for the neutrinos, and so test the model. A neutrino detector requires a large amount of matter for neutrinos to interact with to produce a measurable number of interactions. A 'large amount' in this context is an apartment-sized tank of fluid, a similarly large volume of lead, or a 1km-sized ice cube.

Davis's detector was a tank of 600 tonnes of dry-cleaning fluid, located 1.5km underground at the Homestake Mine in South Dakota. It needed to be there to screen out confounding cosmic-rays signals. Dry-cleaning fluid (tetrachloroethylene) is rich in chlorine atoms, which capture electron neutrinos and transform into radioactive isotopes of argon. Davis analysed the fluid every few weeks to search for argon atoms and found a few tens of them forming in the tank. The sensitivity of Davis's chemical measurements was remarkable enough in itself, but his results were deeply puzzling. It indicated that the flux of solar neutrinos was only about a third of that expected. Were the measurements wrong, was the solar model wrong, or was something else happening?

The discrepancy, the *solar neutrino problem*, was not resolved for three decades. In the interim, many tests and ideas were put forward, including the notion of neutrino oscillations, proposed by Italian and Soviet physicist Bruno Pontecorvo. But repeated experiments doggedly reported the same shortfall in solar neutrinos—two-thirds of them were missing. The breakthrough occurred in 1998, when the Japanese Super-Kamiokande neutrino experiment, a huge underground tank containing 50,000 tonnes of ultra-pure water lined with photomultiplier light detectors, demonstrated the reality of neutrino oscillations. Although the predicted number of *electron* neutrinos are produced in the core of the Sun, by the time they entered Davis's detector, which was sensitive only to electron neutrinos, they had morphed into a mixture of the three flavours. There turned out to be nothing wrong with Davis's measurements or Bahcall's solar model. The fault lay in our understanding of neutrinos.

Neutrinos are not only produced in stars and supernovae, but they are also generated when high-energy cosmic rays, such as protons, smash into the Earth's atmosphere. The cascade of energetic particles decays into others, including GeV-energy atmospheric neutrinos and antineutrinos. It was these atmospheric neutrinos that the Super-Kamiokande detected, and which led to the discovery of neutrino oscillations. The interactions of such high-energy neutrinos in matter produce a plethora of energetic charged particles, moving faster than the speed of light in the water tank. Such particles produce bursts of blue light, a form of Cherenkov radiation, which is analogous to the boom of a shock wave from a supersonic jet aircraft. The characteristic pattern of light picked up by the sensors is back-tracked to infer the energy, position, time, and direction of an incoming neutrino.

High-energy neutrinos are produced in supernovae and the cosmic particle accelerators of AGN as well as tidal disruption events. The energy of neutrinos observed in supernova SN1987A was around 50 MeV. Detecting such bursts provides an early warning system for multi-wavelengths follow-up observations. There is a real-time SuperNova Early Warning System which, on the receipt of a neutrino burst, sends out alerts to astronomers to search for any corresponding electromagnetic or gravitational wave observations. The neutrino energies produced in AGN are, however, 100 billion times higher. To detect these neutrinos, in 2010 a large-volume Cherenkov detector was built out of ice, the IceCube Neutrino Observatory. IceCube is a 1-cubic-km block of ice which, as well as being a neutrino telescope, is an integral part of the ice shelf near the Amundsen–Scott South Pole station in Antarctica, a place where the ice is so transparent that the Cherenkov light can travel hundreds of metres through it to the detectors. IceCube consists of 86 2.5-km-deep holes melted in the ice, down which strings of photomultiplier tubes were lowered.

In 2013, IceCube detected a diffuse isotropic flux of high-energy astrophysical neutrinos and, in 2017, made a major breakthrough in the field by detecting the first very high-energy (300 trillion eV) neutrino coming from a patch of sky in the constellation of Orion. The event, IceCube-170922A, had over 20 times the energy available in the LHC experiment, and triggered a world-wide alert. The Neil Gehrels Swift Observatory identified the event with the flare of a blazar in an AGN some six billion light years away. The Fermi space telescope confirmed that the blazar was emitting gamma rays with energies up to 400 GeV at the time, and other telescopes observed emission across all electromagnetic wavebands. In 2019, IceCube detected another high-energy neutrino which was identified by ZTF (event AT2019dsg) as the fading afterglow of a tidal disruption event of a star being consumed by a 30-million-solar-mass supermassive black hole. In 2022, IceCube reported neutrino observations of a local active galaxy, M77, a spiral containing the radio source Cetus A, powered by accretion onto a supermassive black hole. Over the course of a decade, IceCube has detected some 80 high-energy (teraelectronvolt) neutrinos from the galaxy. However, no gamma-ray counterparts have been seen. This finding underlines the important role played by neutrino observations in observing AGN in which the associated gamma rays are blocked by the gas and dust lying near the nucleus.

Gravitational waves

Light and gravitational waves have one property in common—they both propagate at light speed. However, gravity is a weak force and has only one sign of charge (gravitational masses always add up to produce bigger gravitational effects), whereas electrical forces, whilst being intrinsically much stronger, can be positive or negative, and electrical charges almost completely cancel out in macroscopic pieces of matter. Light propagates as oscillations of the electromagnetic field through spacetime, whereas gravitational waves are oscillations of the fabric of spacetime itself.

Unlike light, gravitational waves are not absorbed or scattered, and can pass unattenuated through clumps of matter on all scales. These properties make gravitational waves a valuable astronomical messenger that brings us information from electromagnetically unobservable regions of the Universe. Gravitational wave sources involve the movements of large amounts of mass and energy, such as in the conditions just after the Big Bang, the mergers of black holes and neutron stars, and the collapse of massive stellar cores in supernovae.

Einstein's General Theory of Relativity describes the curvature of spacetime around massive objects. In a merging binary black hole, the deep pits of gravitational potential energy around each of them coalesce to form a spinning dumbbell structure. As the two black holes spiral inwards, their orbital speeds reach an appreciable fraction of the speed of light, and the disturbances in spacetime propagate outwards as gravitational waves (Figure 32) in a pattern resembling streams of water droplets flung out from a rotating garden sprinkler. Gravitational waves carry away energy from the system. This makes the black holes spiral inwards

32. Gravitational waves spreading out from a pair of inspiralling black holes.

ach other, rotate even faster, and generate even larger gravitational waves.

The binary pulsar

Gravitational waves were in fact first observed indirectly in 1974. American astronomers Russell Hulse and Joe Taylor had been searching for radio pulsars using the Arecibo radio telescope and discovered a pulsar, PSR1913+16, with an odd signal. Instead of regular clock-like pulsations, its pulse rate varied systematically over a period of 7.75 hours. This was the Doppler signature of a pulsar moving at high velocity in a close binary orbit with a massive companion. The pulse rate increases when the pulsar is approaching, and slows when moving away. The companion turned out to be another neutron star of similar mass (about 1.4 solar masses), making this a binary neutron star. The observation of an extremely precise clock (the pulsar) orbiting in the strong gravitational field of a second neutron star provides an opportunity to test General Relativity. The precise pulse timings enabled the astronomers to measure the masses of both neutron stars with a precision better than 0.05 per cent. Over longer timescales, the orbital period was found to be gradually decreasing at precisely the rate that General Relativity predicts for a pair of neutron stars orbiting a common centre of mass and losing energy by gravitational wave radiation. Since the discovery of the binary neutron star, the orbital period has shortened by about a minute, and a merger is predicted in about 300 million years' time.

Detecting gravitational waves

Gravitational waves exert tidal forces on their surroundings by alternately stretching and squeezing space, oscillating perpendicularly to the direction of the wave. Imagine a gravitational wave passing along the axis through a circular ring of test particles floating freely in space (Figrue 33(a)). In the first half-cycle of the wave, the tidal force squeezes the circle

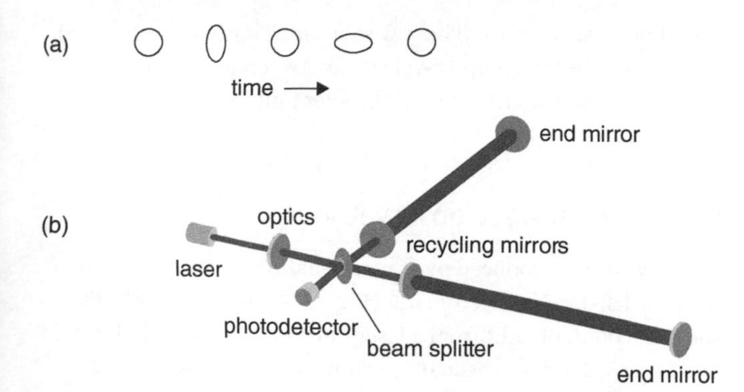

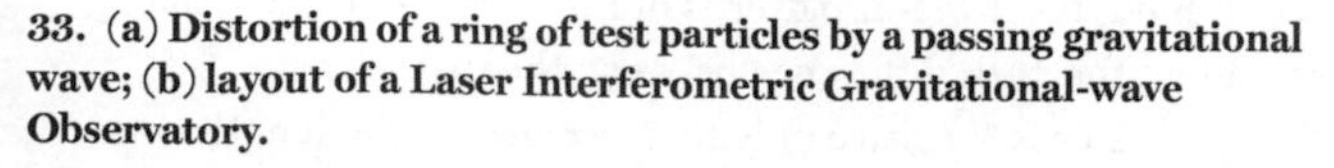

33. (a) Distortion of a ring of test particles by a passing gravitational wave; (b) layout of a Laser Interferometric Gravitational-wave Observatory.

horizontally, and stretches it vertically. In the next half-cycle, the distortions reverse, and then the cycle repeats. Gravitational waves are detected by measuring the amount of geometrical distortion they produce. This is quantified by the *strain*, the fractional amount by which a wave alters distances. Just like the amplitude of an electromagnetic wave, the strain of a gravitational wave drops off inversely with distance (it is the power of the wave that falls inversely with the square of the distance).

Merging binary black holes produce tiny but measurable gravitational wave signals at cosmological distances. A test object placed near to a pair of merging 10-solar-mass black holes, for example, would experience a strain of about unity; that is, the object would be stretched and squeezed by an amount equal to its size. But at a distance of a billion light years, the strain drops to around 10^{-21}. Such a small strain has the effect of displacing two test points, separated 4km apart, by about 1/200th the radius of a proton. One might imagine that such a tiny distance could, perhaps with great care, be measured by some sort of sophisticated ruler. However, any physical object is itself affected by the same stretchings and squeezings as the test points, and so ould not 'see' the wave. The way such small strains are detected is by measuring the light travel time between test points, in effect converting a *distance* measurement into a *time* measurement.

Gravitational wave observatories

The tiny strains produced by gravitational waves are measured with a Michelson interferometer (Figure 33(b)). This works by splitting coherent light from a laser into two beams, sent along two L-shaped arms, reflected by end mirrors, and then recombined, resulting in patterns of interference fringes that depend on the path difference between the two. The interferometer is sensitive only to *differences* in the lengths between the two arms so that when a gravitational wave passes by these differences show up as changes to the interference fringe patterns measured at the photodetector.

The world's first gravitational wave telescope, the Laser Interferometer Gravitational-wave Observatory (LIGO), started operating in 2015 at two different sites. With detectors of such extreme sensitivity, the risk of false triggers is high, and any valid detections must also be seen at other sites. The two US interferometers are at Livingston, Louisiana, and in Hanford, Washington, 3,000km away. Each has 4km-long arms and, at the end of these, masses are suspended on complex anti-seismic-vibration mounts carrying mirrors. A third mass in the centre supports a beam splitter which reflects half the laser beam into one arm, and transmits half into the other. The effective optical length is considerably larger than 4km because the laser beams are made to bounce back and forth hundreds of times using recycling mirrors. The beams are eventually returned to the splitter which again half transmits and half reflects the light, so that part of each beam's light is combined and returned to the laser, and part is sent to the photodetector to form interference fringes. While the basic operating principle of a LIGO is relatively straightforward, making a practical instrument with the required level of sensitivity was a major challenge. Numerous practical measures were implemented to reduce extraneous noise, including operating the entire interferometer in a high vacuum to avoid light scattering as well as the vibrations caused by the pummelling of air molecules on the mirrors.

The first gravitational wave ever detected directly (GW150914) came as a 'chirp' lasting just half a second at Livingston on 14 September 2015 (Figure 34). A near-identical signal was seen seven milliseconds later at the Hanford site, corresponding to the light travel time between the two sites. The 'chirp' has a distinctive pattern, beginning as a regular weak, low-frequency oscillation at about 200Hz. Gradually the period shortened and the amplitude increased as two black holes merged.

The precise shapes of the waveforms contain a great deal of information about the merger. What is this information and how is it decoded? In parallel with the experiments, a major theoretical effort was devoted to solving Einstein's equations for binary black hole and neutron star mergers. This is a problem in the non-linear dynamics of curved spacetime (named *geometrodynamics* by the American physicist John Wheeler) and is analytically intractable. Fortunately, progress was made computationally, and a suite of numerical simulations has been generated using a supercomputer to produce a library of millions of signatures of combinations of mergers. These can then be compared with data. The best-fitting model to GW150914 (the thin line in Figure 34) indicates the merger of two black holes, weighing 29 and 36 solar masses, to one black hole weighing 62 solar masses. The mass deficit—three solar masses—was radiated away as gravitational wave energy. The waveforms contain information on the absolute gravitational wave *luminosity* of the merger which meant that, given the strain amplitudes measured by LIGO, it was possible to infer the gravitational wave luminosity distance of the merger. For the GW150914 event this was 1.3 billion light years.

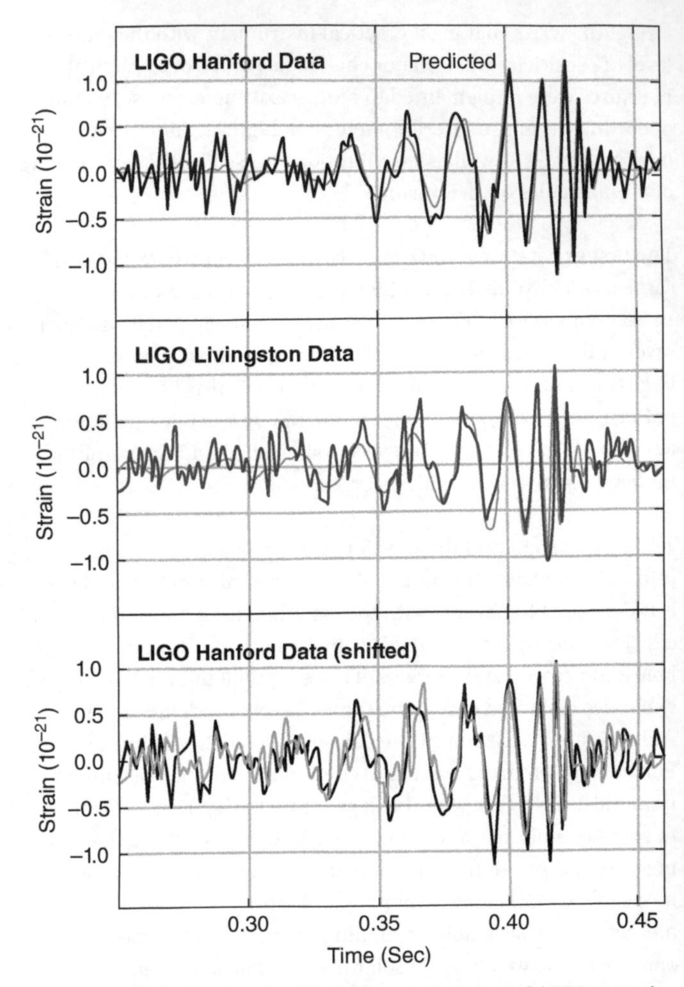

34. Traces of the first gravitational waves ever detected (GW150914), from merging black holes, at the two LIGO observatories. The top plots show the outputs at the individual sites, together with predictions (thin lines) of the signals according to Einstein's General Relativity. The bottom trace compares the signals after correcting for the seven-millisecond time delay it took for the signal to reach the Hanford site.

When two black holes merge there are three distinct phases. The first is an 'inspiral' phase, during which energy is lost to gravitational radiation, and is characterized by oscillations of increasing frequency and amplitude as the black holes spiral inwards. In the second, 'merger', phase, the amplitudes and frequencies peak as the event horizons touch and begin to merge. In the third and final 'ringdown' phase, the resulting bigger black hole rings like a bell as the distortions on the event horizon are smoothed out, and the oscillations die away.

A single LIGO is virtually an all-sky detector, but one that has four 'blind spots' which arise, for example, when a gravitational wave arrives along a line bisecting the two arms, in which case the arms are stretched and squeezed by equal amounts and no output is produced. With a two-interferometer measurement, some positional information about the source can be retrieved. For example, the GW150914 signal was localized to an area of 600 square degrees in the Southern Hemisphere, equivalent to 3,000 times the size of the full Moon. Adding more LIGO detectors improves precision, allowing source positions to be better triangulated. In 2017, a third interferometer, Virgo, started operating near Pisa in Italy. With three LIGO/Virgo detectors, the error box within which there is a high probability of a gravitational wave source in the sky shrinks to typically a few hundred square degrees. A fourth detector, the underground Kamioka Gravitational Wave Detector (KAGRA) in Japan, came online in 2021.

The present generation of ground-based LIGOs are sensitive to wave frequencies in the range 20–2,000Hz which corresponds to signals from stellar mass black hole and neutron star mergers, inspiralling from a separation of a few Schwarzschild radii. To date, around 90 black hole mergers have been detected. These show a wide diversity of black holes weighing from 5 to almost 150 solar masses, considerably larger than what was known or expected. Future gravitational wave observations will also be made m space, which offers two clear advantages. First, space offers a quiet, noise-free environment in which the test masses and mirrors of a Michelson interferometer float weightlessly in their individual satellites, thus avoiding Earth's seismic activity which affects gravitational wave detectors at frequencies lower than 10Hz. In space the interferometer arm lengths can be extended to a few million kilometres, which would mean that such an interferometer would be sensitive to medium-frequency signals (between 0.0001Hz and 1Hz). Such a telescope would be able to detect close-spaced white dwarf binaries in the Milky Way, track binary stellar-mass black holes many months before coalescence, detect the mergers of supermassive black holes in galaxies at cosmological distances, and even observe a putative gravitational wave background radiation emitted in the early Universe. In 2017, the proposed Laser Interferometer Space Antenna (LISA) space probe settled on a design for three spacecraft orbiting at the Lagrangian L2 point, spaced apart by 2.5 million km, and arranged in an equilateral triangle. The LISA pathfinder experiment, launched in 2015, has successfully validated the technologies needed for the full LISA system.

Kilonova

A merger between two isolated black holes generates gravitational waves but emits little or no electromagnetic radiation. However, the collision of two neutron stars produces both gravitational and electromagnetic wave signals. In 2017, the three interferometers of the LIGO–Virgo collaboration observed a 30-second-long chirp from the first binary neutron star merger (GW170817) ever observed. It was localized to a region of the sky in which the Fermi space telescope observed a short gamma ray burst, 1.7 seconds later. The optical counterpart was located to a spot of light in the outskirts of a lens-shaped galaxy, NGC4993, 130 million light years away and was discovered by the 1m Henrietta Swope optical telescope at the Las Campanas observatory in Chile. This identification precipitated a spate of follow-up observations using many telescopes across the globe. These showed GW170817 to have the signature of a *kilonova*—a flare 1,000 times more luminous than a nova, or 100 million times the luminosity of the Sun. The fireball of a kilonova is powered by the radioactive decay of heavy elements produced in the merger. Within weeks, the afterglow became visible first in X-rays and then at radio wavelengths, reaching its peak brightness six months later; it was still visible at 1,000 days. No neutrinos were detected by the IceCube observatory.

The general picture is that the GW170817 event resulted from the inspiral and collision of two neutron stars, smashing into each other at around a third of the speed of light, and spewing out vast chunks of hot nuclear matter into space. The expanding cloud of nucleon-rich debris consisted of heavy nuclei which were immersed in an intense flux of energetic neutrons. The nuclei captured neutrons (via the nuclear rapid neutron-capture process, known as the *r-process*) to form unstable highly radioactive neutron-rich nuclei. These nuclei decayed relatively slowly to form heavy elements and released energy which ultimately lit up the fireball. The hot kilonova shell, bearing a rich cargo of some of the heaviest elements in the periodic table, then ploughed into the interstellar medium. This picture is supported by spectroscopic observations from the VLT which observed spectral lines of heavy rare earth elements (lanthanides) in the afterglow, indicating the presence of periodic table chemical elements from niobium to uranium, and supporting the theory that heavy elements are produced in binary neutron star collisions. Astronomers have estimated that the collision giving rise to GW170817 ejected several Earth masses of gold and platinum into space.

Towards the end of the chirping signal from GW170817, the wave signal became lost in the noise, which meant that any ringdown phase was not observed and so the final state of this event is unknown. With future higher sensitivity gravitational wave observations, astronomers hope to be able to piece together a more complete picture of the nature of binary neutron star collisions, the role of kilonovae in enriching the chemistry of the interstellar medium, and the structure of neutron stars. In addition, with GW170817, the near-simultaneous arrival of the electromagnetic and gravitational waves showed their speeds to be the same to within one part in a quadrillion. The event also showed that the short gamma-ray burst was clearly associated with the kilonova.

The optical signals associated with GW170817 were fast transients, peaking in brightness within one day and starting to fade after two. This feature, allied with the large error boxes for gravitational wave signals on the sky, makes it challenging to locate the optical counterparts of other similar neutron star collisions. In 2022, a new pair of robotic telescopes, the Gravitational-wave Optical Transient Observer (GOTO), started operating at the Roque de Los Muchachos Observatory in La Palma in the Canary Islands. GOTO is designed to scan the skies rapidly for any fast-evolving optical counterparts on the receipt of a gravitational wave trigger. Each system uses an array of eight 0.4m, f/2.5 telescopes to achieve a large overall field of view and, in its search for optical transients, can cover large areas of the sky rapidly.

An exciting prospect for future gravitational wave observations is the possibility that compact object mergers could serve as what are known as *standard sirens* and be used to measure the Hubble constant. This was suggested by American physicist Bernard Schutz, and offers a completely different way of measuring the constant, because it relies on the self-calibrating property of a gravitational wave signal: the information carried by the gravitational wave encodes the distance of the source. For mergers with associated electromagnetic counterparts, such as binary neutron star collisions, optical telescopes can be used to measure the corresponding redshift. Such multi-messenger observations would complement each other, since gravitational wave observations are good for measuring distances, whereas electromagnetic wave observations are good for measuring redshifts. The attraction for cosmology is the possibility of being able to construct a new type of Hubble diagram, neatly sidestepping the present uncertainties in the distance measurements.

Chapter 7
A bigger picture

The close symbiotic relationship between observational astronomy and astrophysics is inspiring new ideas and models, and these, in turn, suggest more penetrating observations. In this final chapter, let us consider how observations have shaped our knowledge of the structure of the Universe as a whole, which is embodied in the so-called standard model of cosmology, the lambda-cold-dark-matter model (see below). Its theoretical structure is underpinned by the huge intellectual achievement of Einstein's General Theory of Relativity, worthy of a law of the Universe in its own right. The two key observational discoveries supporting the model are of the expansion of the Universe, and the CMB radiation.

Newton's theory of gravitation successfully accounts for the detailed motions of all the planets in the solar system except one. In 1855 French astronomer Urbain Le Verrier pointed out that Mercury, the planet nearest to the Sun, showed an anomalous orbital precession which could not be explained by Newtonian theory. The perihelion of an orbit is its closest point to the Sun and, when an orbit precesses, the perihelion rotates around the star. Einstein's first test of general relativity used these observations to account successfully for Mercury's motion. As Mercury approaches its perihelion, it moves deeper into the Sun's gravitational potential well. Its motion in this region of more curved spacetime causes the perihelion to advance exactly as predicted by Einstein's theory.

Bolstered by this success, in 1917 Einstein applied his equations to model the Universe as a whole. In this, he assumed the *cosmological principle*, namely that the Universe has no preferred centre and, on large enough scales, looks the same in any direction. The model predicts three possible geometries of the Universe, depending on how much matter is present. If the matter density is large, space curves to form a closed or spherical Universe in which angles of a triangle in space add up to more than 180°, and one in which the expansion will eventually slow down and be followed by contraction towards a big crunch. If the matter density is small, space curves into a saddle shape (open Universe) in which the angles of a triangle add up to less than 180°, and the Universe will continue to expand forever, ending ultimately in what has been called the big freeze. Between those two cases is the finely balanced geometry of a flat (Euclidean) Universe where the angles of a triangle add up to 180°, and where the Universe contains just enough matter to slow the expansion down.

When Einstein published his model, the Universe was believed to be static (the discovery of its expansion was still a decade away). And this brought a problem—the matter in Einstein's model Universe kept on moving around. The basic instability goes back to Newton's conjecture of a static but delicately balanced Universe consisting of an infinite cubic lattice of mass points. The attractive gravitational force on each mass due to the others cancels exactly by symmetry. However, the equilibrium is precarious, for if any mass were to be displaced, even slightly, the unbalanced forces lead to a wave of collapse which spreads throughout the whole system. The problem is akin to trying to balance a pencil vertically on its point. The slightest nudge from an errant air molecule will eventually cause it to topple.

Einstein discovered that his model could, however, produce a static Universe if he added an extra term to the field equations, the *cosmological constant*, denoted by the Greek letter, lambda. The lambda term has the effect of pushing back on spacetime, opposing the gravitational attraction between masses, and effectively prevented them from moving around. But, within a few years, Edwin Hubble's observations showed that the Universe was in fact expanding, and Einstein regretted including the lambda term. He remarked that the cosmological constant had been his 'greatest blunder'.

In the 1920s, Russian physicist Alexander Friedmann and George Lemaître independently discovered solutions to Einstein's field equations that described evolving Universes. Lemaître extrapolated the solutions back in time to a state of infinite density and temperature, which he envisaged as coming from the explosion of a 'primeval atom'. This was the 'Big Bang', a term coined derisively by English astronomer Fred Hoyle. Hubble's observation of the recession of the galaxies supported the Big Bang model, but many scientists remained sceptical because it implied that the laws of physics would break down at early times. The most prominent rival theory was the Steady-State model of Hermann Bondi, Thomas Gold, and Hoyle who in 1948 proposed that as space expanded, matter was being created continuously at just the rate needed to compensate for the expansion. This implied the Universe has always looked the same and there was no moment of creation. By the early 1960s, the radio observations of Martin Ryle and colleagues argued against the Steady-State model by revealing an excess density of radio galaxies at large redshifts, implying that the Universe had undergone strong evolution and therefore has not always looked the same. At that point in the story, however, new observational evidence turned up that clinched the Big Bang.

The cosmic microwave background radiation

In 1965, two American scientists at Bell Labs were experimenting with an old, oddly shaped microwave antenna in which a couple of pigeons had made a home. The scientists had no idea that what they were about to discover with it would shake the foundations of cosmology. The 15m Holmdel horn antenna, resembling a giant ear trumpet, had once been used to bounce signals off orbiting metallized balloons, a sort of early cellphone network originally proposed by the English science-fiction writer and futurist Arthur C. Clarke. The scientists wanted to repurpose it as a radio telescope. They tested it by first pointing it at a blank piece of sky. But it did not perform well—the noise level in their receiver was about 100 times bigger than it should have been. Painstakingly they eliminated sources of interference from broadcast and radar sources, and even drove away the nesting birds, scraping off the droppings coating the inside of the antenna. But it made no difference, the noise remained, and it seemed to be coming equally from everywhere in the sky. After eliminating every conceivable source of interference, they concluded that the noise must be coming from the cosmos. They had stumbled on the *cosmic microwave background radiation* and, for that discovery, the scientists, Arno Penzias and Robert Wilson, were awarded the 1978 Nobel Prize.

Why was this pervasive radio hiss important? After Lemaître had published his primeval atom theory in 1927, a Russian Ukrainian theorist working in the US, George Gamow, and colleagues, started looking at what is now called the Big Bang model of the Universe. In 1948 they argued that such a model correctly predicts the primordial abundances of hydrogen and helium (baryons) in the early Universe. Importantly, the theory also predicted that the Universe should now be filled with a faint glow of radiation, the fossil remnant of thermal radiation from the primordial fireball.

The early high-density Universe was a searingly hot plasma of baryons, electrons, photons, and neutrinos. Photons scattered frequently from free electrons which, in turn, were strongly coupled to the baryons. The photons and matter formed a tightly coupled baryon-radiation fluid, in which the radiation was trapped, rendering the plasma opaque, like light trapped in a dense fog. As the Universe expanded it cooled, and, at a temperature of around 3,000K, it was energetically favourable for the electrons to bind to the protons to make hydrogen atoms. At that point the photon scattering ceased, and the fog cleared. This phase-change occurred when the Universe was 380,000 years old, marking the *epoch of decoupling*, the time when the light was first free to travel large distances across the cosmos. The boundary surface which separates the opaque phase from the transparent phase has a redshift of about 1,000 and is known as the *last scattering surface.*

The CMB radiation was emitted from the last scattering surface 13.8 billion years ago and enters our telescopes from a spherical surface around us, making the CMB photons the oldest that can be observed. After the radiation became decoupled from matter, the expansion of space stretched the CMB waves from visible to microwave wavelengths. Therefore, when we now observe it, 13.8 billion years later, we see a bath of thermal radiation that is 1,000-fold cooler (it is now close to 3K) than its temperature at the last scattering surface. Since the radiation has had no interactions with matter since it left the last scattering surface, it has retained its thermal spectrum. This spectrum was confirmed by the Cosmic Background Explorer (COBE) satellite in 1990. In fact, the spectrum has the most precise blackbody form that has ever been measured. Overnight, COBE's observations changed cosmology from an approximate science into a high-precision one.

The brightness of the CMB is almost constant over the sky. But it is not perfectly smooth, and its tiny temperature irregularities contain information. At the level of one part in 1,000 there is a smoothly varying distortion called the *dipole asymmetry*. This tells us about our motion through space. The CMB photons coming towards us appear blueshifted and those travelling away from us are redshifted. At an even lower level of one part in 100,000, there is a finely woven tapestry of temperature ripples printed on the sky. These ripples were first seen by COBE and were later mapped with higher sensitivity and resolution by NASA's Wilkinson Microwave Anisotropy Probe (WMAP) in 2006 and by ESA's Planck spacecraft in 2015 (Figure 35). The brighter hot spots are regions where the radiation is more intense, and the cold spots are where it is less so. The difference between the peaks and troughs is incredibly small.

The tiny temperature fluctuations can be understood by considering the conditions in the hot baryon-radiation fluid prior to the epoch of decoupling. Here there were two competing forces: gravity and radiation pressure. Imagine a clump of fluid with a higher-than-average density. Gravity enhances the density in the clump by attracting more matter into it from its surroundings. But the photons in the clump now scatter more frequently and exert

35. The whole-sky map of temperature fluctuations in the cosmic microwave background radiation measured by the Planck space mission. It is a window on the early Universe.

an increased outward pressure trying to push the clump apart. If the clump is larger than a critical size, gravity wins and the density in the clump increases. If the clump is small the radiation pressure provides a restoring force, pushing back on the clump and setting up sound wave oscillations, which propagate as longitudinal compressions and rarefactions in the fluid at speeds close to the speed of light. These primordial sound waves are called *Baryon Acoustic Oscillations* (BAOs). BAOs resemble the standing-wave vibrations of air in an organ pipe or a guitar string, and are similarly characterized by amplitudes, fundamental frequencies, and harmonics. The early Universe hummed like a musical instrument. But when radiation and matter became decoupled, the BAOs died out, leaving frozen-in patterns imprinted on the last surface of scattering, traces preserved in the fossil radiation.

The CMB fluctuations show us the very earliest seeds of all the structures of galaxies and clusters of galaxies we now see in the cosmos. The map contains a vast amount of information such as the age of the Universe, what its beginning was like, its composition, geometry, expansion rate, and when the first stars lit up. A quantitative way to compare this information with cosmological models is to study the power spectrum, that is, how much the temperature fluctuations vary and are correlated over different angular scales on the sky (Figure 36). An angle measure on the sky can be related to a linear distance on the last scattering surface. The spectrum therefore tells us about what types of acoustic waves were present at this time. The most prominent feature in the spectrum is the first acoustic peak, which has an angular size of about 1°, which indicates that the Universe has a flat Euclidean geometry.

Inflation

Although the Big Bang model successfully accounted for the origin of the light elements and the thermal Planck spectrum of

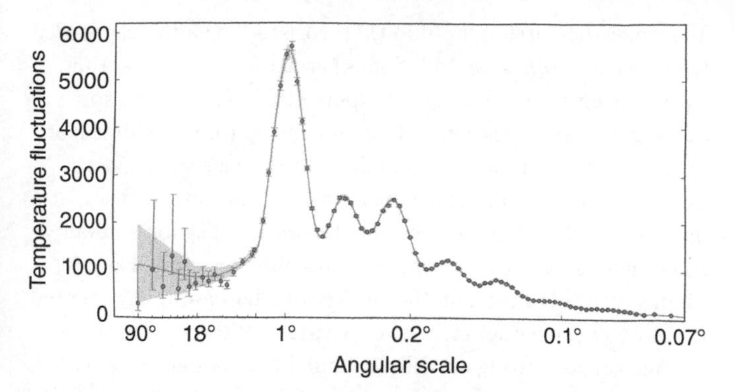

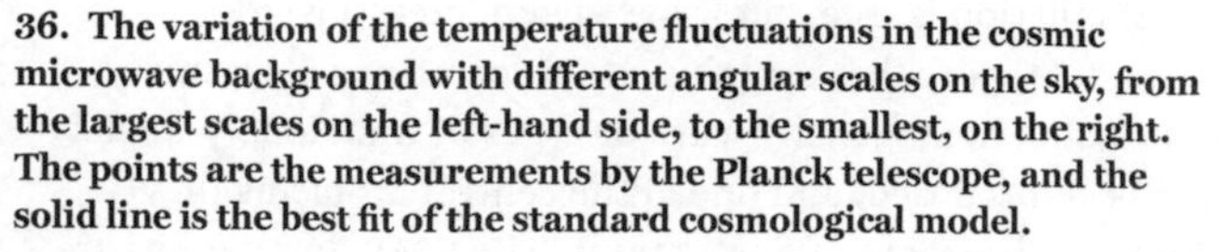

36. The variation of the temperature fluctuations in the cosmic microwave background with different angular scales on the sky, from the largest scales on the left-hand side, to the smallest, on the right. The points are the measurements by the Planck telescope, and the solid line is the best fit of the standard cosmological model.

the CMB, it also had some basic inconsistencies. These included the *flatness problem* (why is the Universe observed to have a finely tuned flat geometry), and the *horizon problem* (why does the Universe look so homogeneous). For example, when we measure the temperature of the CMB in one direction in space, it is very close (to within one part in 10,000) to that measured in the opposite direction. These two regions are physically so far apart that if the expansion of the Universe is run backwards in time, the basic Big Bang model would not have allowed them to have been in causal contact with each other. Another way of looking at this is to ask what the maximum distance is that light could have travelled on the last scattering surface since the Big Bang, a distance called the *particle horizon length*. The angle corresponding to the particle horizon length in the power spectrum in Figure 36 is about 1.7°, which tells us that scales larger than that were not then in causal contact. However, this basic fact seems to conflict with the observed high degree of uniformity of the CMB across the sky, which suggests that all scales were in fact once in causal contact.

To resolve these issues, in 1980 US theorist Alan Guth proposed the concept of *inflation*. Inflation is based on the idea that the early Universe went through an exponentially fast growth spurt during the first 10^{-32} seconds, when it blew up from a billionth of the diameter of a proton to something the size a grapefruit, an increase of about 20 orders of magnitude. This happened so quickly that the inhomogeneities in the matter distribution were smoothed out, and the curvature of spacetime expanded away. Picture yourself sitting on the surface of a balloon being inflated rapidly to a huge diameter—the curvature of the part of the surface you are sitting on will very quickly look flat. We have no idea if inflation is true, but it has engendered the most successful theory that explains the observations of cosmic flatness and homogeneity. Various types of inflationary models have been proposed, and one of the central problems they address is the origin of fluctuations. The CMB fluctuations are believed to have arisen from microscopic quantum fluctuations which, during the period of inflation, became stretched to cosmic scales.

Dark matter

After Zwicky's observations of galaxy clusters and his inference about the existence of dark matter, the reality of dark matter became more firmly established in the late 1970s when US astronomer Vera Rubin made detailed spectroscopic observations of the rotational velocities of the stars and gas clouds in the outer fringes of spiral galaxies, such as the Andromeda galaxy. She found that the galaxies were rotating too quickly for the amount of luminous visible matter they contain. If the dominant mass in galaxies resided only in the stellar population, it was argued, the stars should all be flung off into space. Rubin calculated that there must be about five times as much dark matter present as visible matter to bind galaxies together. The dark matter is believed to be distributed in diffuse halos around the visible galaxies. The view now is that the luminous baryonic matter in stars and galaxies in the Universe should be regarded as an impurity in which the dominant gravitational mass (now estimated to be around 85 per cent of the total) is invisible.

While there is abundant observational evidence for dark matter, nobody has yet identified what it is, and it remains one of the biggest mysteries. Dark matter has the properties of being electrically neutral and interacting mostly with itself or with normal matter through gravity. It has been conjectured that it could possibly be explained by a population of small, unobserved black holes. However, the current leading hypothesis is that dark matter is a new, so-far unseen, kind of elementary particle, a *weakly interacting massive particle* (WIMP) which, apart from gravity, interacts only via the weak nuclear interaction. Physicists know of no particle that resembles this description in the Standard Model of Particle Physics, the best model of fundamental matter and forces we currently have. Despite many searches through the data coming from the highest energy accelerators like the LHC, no viable candidate subatomic particle with the required properties has yet been observed.

It is possible that a population of WIMPs could have formed in the hot dense plasma of the very early Universe, and have survived to the present. Estimates of the interaction of WIMPs with ordinary matter via the weak interaction are not zero, just very small, a possibility that has inspired experiments to detect them. One, the LUX-ZEPPELIN dark matter experiment located 1 mile underground at the Sanford research facility in South Dakota, consists of a tank of 7 tonnes of liquid xenon, surrounded by banks of light detectors. The idea is to search for very rare interactions between the hypothetical WIMPs and heavy xenon nuclei which could produce flashes of light and free electrons. Although no detections have been reported so far, the experiments have ruled out many combinations of dark-matter–normal-matter interaction strengths and WIMP masses.

An alternative possibility, conjectured by US theorists Steven Weinberg and Frank Wilczek, is that dark matter consists of hypothetical particles called *axions*. Axions have interesting properties: they interact weakly with themselves and with normal matter, have small masses, are neutral, and could have been formed in abundance in the Big Bang. Axions have a unique property of transforming to photons in the presence of a very strong magnetic field. Experiments like CERN's axion solar telescope (CAST) are searching for solar axions that should convert to X-rays inside a powerful superconducting magnet. So far, searches for axions have drawn a blank.

Dark energy

In 1998, two groups of US astronomers set out to measure the Hubble constant over cosmologically large distances using Type Ia supernovae as standard candles. The astronomers were shocked to observe that the supernovae were systematically fainter (and therefore further away) than had been expected from the Hubble law. This implied that the space through which the light had travelled had expanded more than had been expected. The shock lay in the fact that the expansion of the Universe appears to be *accelerating*, which is not what is expected in a Universe filled with gravitationally attracting masses which will generally tend to slow down the expansion. The astronomers had found evidence of 'something new' pushing galaxies apart. The degree of surprise was (and still is) akin to tossing a ball up into the air which, instead of returning to Earth, accelerates away and keeps on going.

An acceleration in the cosmic expansion rate is, however, exactly what Einstein's lambda term can provide, appearing as a kind of negative gravity, or *dark-energy*-filling space. Dark energy has the property that its *density* is constant, which means that the amount of it increases in proportion to the volume created by the expansion of the Universe. At the present epoch the mass-energy of dark energy is tiny. For example, inside the volume of the Earth it is equivalent to a few millionths of a gram. But, summed over vast intergalactic volumes of space, it dominates the mass-energy content of the Universe.

Dark energy presents us with an enigma. If we think about the most perfect of vacuums, a region of space from which all the particles and photons have been removed, there is still something present—*vacuum energy*. Vacuum energy is a prediction of quantum field theory which tells us that, at the quantum scale, there is an intense flurry of activity where virtual quantum particles continually flit in and out of existence contributing to a background energy density. The existence of vacuum energy has been confirmed in experiments on the *Casimir effect*. This is a measurable force acting on two parallel plates arising from vacuum quantum waves pushing them together. In macroscopic terms, imagine two boats lying close to each other and side by side in a choppy sea. The wave pressure pushing on the outsides of the hulls is larger than that pushing them apart, because there is not enough space between the hulls for the opposing long-wavelength waves to fit. This imbalance results in a net inward force. It is tempting to try to identify the cosmological lambda with the energy of the vacuum. However, this engenders a serious problem, pointed out by the Soviet theorist Yakov Zel'dovich in 1967. By adding up the energy of all the virtual quantum particles, he deduced that the energy density is over 100 orders of magnitude too big to explain the lambda term in the standard model of cosmology. This problem has been dubbed the worst discrepancy in all of science.

The standard model of cosmology

Looking at the bigger picture, the present status is this. We possess two snapshots of the Universe: A and B. Snapshot A is a picture of the Universe in its infancy, the CMB at an age of 380,000 years (Figure 35). Snapshot B is the cosmic web in the local Universe coming from galaxy surveys (Figure 8). Just as the pattern of bright lights which decorate a Christmas tree, seen on a dark night in the distance, lead us to infer the existence of the tree itself, observations of the spatial distribution of the luminous galaxies inform on the underlying structure of the cosmic web. While snapshot A shows the infant Universe to be extremely smooth (to one part in 10,000), snapshot B, taken almost 14 billion years later, shows the distribution of matter to have become very lumpy. One of the central tasks of modern cosmology is to explain how state B arose from state A. Small fluctuations in the density of matter grow bigger over time via the mechanism of *gravitational instability*, which causes over-dense regions of the Universe to grow bigger by drawing in matter from adjoining under-dense regions.

The standard model of cosmology is a mathematical model based on the laws of physics including general relativity, classical mechanics, thermodynamics, electromagnetism, quantum mechanics, and atomic and nuclear physics. This complex model is analytically intractable but, by the 1980s, computers became powerful enough to simulate large-scale cosmic structure by observing how ensembles of test particles interact, evolve, and clump under gravity. Starting from initially smooth distributions of matter, gravitational instabilities do indeed produce large-scale structures; but, at first, these did not match the cosmic web. It was found that more realistic simulations of galaxies and clusters of galaxies could result if the dark matter is assumed to be cold, which means that it moves around at speeds much smaller than the speed of light. Hence the name *cold dark matter* (CDM). If the dark matter were hot, the clumps would have dissipated, and the observed galaxy structures could not have formed. With CDM included, the models began to produce the kind of structures seen in surveys, but they predicted the Universe to be only about seven billion years old, conflicting with the 10-billion-year-old stars observed in globular clusters.

The simulations could explain the observations better if Einstein's cosmological constant, lambda, was added to the CDM model, producing the *lambda-CDM model.* This solves the modeller's problems with the simulations, but resurrecting the cosmological constant was anathema to many astronomers because it implied that the geometry of the Universe must be flat and finely tuned, which seemed implausible. However, when the COBE CMB observations were announced in 1992, the lambda-CDM model gained support and, when the 1° peak in the CMB power spectrum was discovered, the case became even stronger. The observations of the accelerated expansion then really clinched the case for the model.

The key components of the lambda-CDM model are: Einstein's General Theory of Relativity, the cosmological principle, and the laws of physics, as well as inflation and the dark matter and dark energy. It is the simplest model that predicts a flat Euclidean geometry, and one that fits the peaks in the power spectrum of the CMB fluctuations. (The solid line in Figure 36 is the model prediction which fits the Planck CMB data extremely well.) It also explains the observed large-scale structure in the local Universe, the primordial abundances of the light elements of hydrogen, deuterium, helium, and lithium produced in the hot Big Bang, and the acceleration of the expansion of the Universe. In travelling through 13.8 billion years of cosmic history, the paths of CMB photons get deflected on their way to us because of gravitational lensing by the intervening clumps of matter. Observations of gravitational lensing by the Dark Energy Survey Blanco 4m telescope at Cerro Tololo Inter-American Observatory, in the Chilean Andes, is currently providing information on how the relative influence between dark matter and dark energy has changed over the lifetime of the Universe. Further constraints on the model come from galaxy surveys which show the faint imprints of baryonic acoustic oscillations on the large-scale distributions of galaxies on length scales of around 500 million light years. The best fitting model tells us that the Universe is 13.8 billion years old, has a flat geometry, and consists of 68 per cent dark energy and 27 per cent dark matter, with just 5 per cent in the familiar form of normal atoms, quarks, gluons, and leptons.

The history of the Universe is portrayed as a spacetime diagram in Figure 37, tracing its evolution from the Big Bang to the present, with a notional size scale indicated by the grid. The early quantum fluctuation era is followed by a brief period of inflationary growth. For the next 380,000 years, the cosmic fireball cooled as the coupled baryon-radiation fluid expanded. Then, the radiation decoupled from matter leaving the imprint of the CMB fluctuations on the last scattering surface, before spreading out through the Universe as the afterglow light pattern. The Universe, composed mainly of primordial hydrogen and helium as well as photons, neutrinos, and dark matter, continued its cooling and expansion and entered the cosmic *dark ages*, a time before stars and galaxies had formed.

Dark matter is conjectured to have clumped together in the early Universe, to form a network of filaments, the cosmic web. These filaments are the scaffolding upon which the first stars and galaxies formed. Primordial hydrogen gas is thought to have flowed into this skeletal backbone, where it aggregated in the intersecting nodes of the network in dense hydrogen-rich clouds which collapsed to form the first massive stars. These first stars formed when the Universe was around 400 million years old, and were made almost entirely of hydrogen and helium. This composition enabled them to reach much larger masses (up to 1,000 solar masses) than is possible from the more metal-rich star-forming material available in the local Universe. In the era of reionization, the first stars shone brightly with UV light, ionizing the surrounding gas, and ushering in the cosmic dawn. The massive stars, however, were unstable and exploded as supernovae, injecting the first heavy elements into the interstellar medium, an enrichment that would be recycled into later

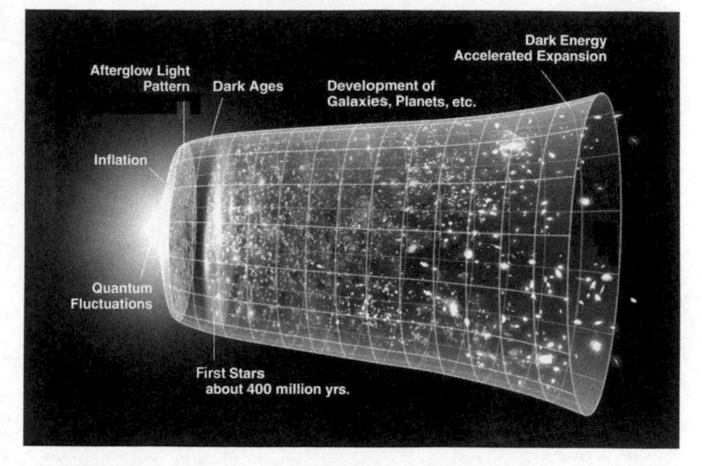

37. Sketch of the timeline of the Universe showing the different epochs in its evolution, from the Big Bang (left-hand side) to the present day (right).

generations of stars. Such massive stars have extremely short lives (of less than a million years) and, when they run out of fuel, they are likely to collapse promptly to form black holes of 1,000 solar masses. It is possible that these first large black holes could be the seeds for the billion-solar-mass black holes we know to have formed remarkably early in the evolution of the Universe.

For the next several billion years the expansion of the Universe gradually slowed as gravity drew the matter together, the stars and gas condensed into galaxies, and stars acquired their planetary systems. As more space was created by the cosmic expansion, the dark energy, which hitherto had played a small overall role, became increasingly dominant and, in the past few billion years, has speeded up the expansion of the Universe.

We have reached the end of this *Very Short Introduction*, and there remain many open questions. The big questions are: what is dark matter? what is dark energy? and what is inflation? There is also another question: are we alone in the Universe? It has been said that whatever the answer turns out to be it will be an interesting one, and will forever change the way that we regard ourselves. In the last few decades, the long-sought-after dream of astronomers to discover exoplanets has been amply fulfilled; the observations have shown us that there are more exoplanets than there are stars, a discovery that has brought us a step closer to answering the question about life in the Universe. The next generation of powerful telescopes may well go further and resolve the question. Above all, the history of observational astronomy is littered with accidental discoveries. We must therefore keep our minds open and be prepared for surprises.

Further reading

James Binney, *Astrophysics: A Very Short Introduction* (Oxford University Press, 2015)
Katherine Blundell, *Black Holes: A Very Short Introduction* (Oxford University Press, 2015)
Pedro Ferreira, *The Perfect Theory* (Little, Brown, 2014)
Francis Graham-Smith, *Eyes on the Sky* (Oxford University Press, 2016)
John Gribbin, *Galaxies: A Very Short Introduction* (Oxford University Press, 2008)
Andrew King, *Stars: A Very Short Introduction* (Oxford University Press, 2012)
Chris Lintott, *The Crowd & the Cosmos* (Oxford University Press, 2019)
Malcom Longair, *The Cosmic Century* (Cambridge, 2013)
Ian McClean, *Electronic Imaging in Astronomy* (2nd edition, Springer, 2015)
W. Patrick McCray, *Giant Telescopes* (Harvard University Press, 2006)
Govert Schilling, *Ripples in Spacetime* (2nd edition, The Belknap Press of Harvard University Press, 2019)
Kip Thorne, *Black Holes and Time Warps* (W.W. Norton, 1994)

Index

For the benefit of digital users, indexed terms that span two pages (e.g., 52–53) may, on occasion, appear on only one of those pages.

C

D

E

F

G

H

I

J

K

L

M

N

O

P

Q

R

S

T

U

V

W

X

Y

Z

GALAXIES

A Very Short Introduction

John Gribbin

Galaxies are the building blocks of the Universe: standing like islands in space, each is made up of many hundreds of millions of stars in which the chemical elements are made, around which planets form, and where on at least one of those planets intelligent life has emerged. In this *Very Short Introduction*, renowned science writer John Gribbin describes the extraordinary things that astronomers are learning about galaxies, and explains how this can shed light on the origins and structure of the Universe.

www.oup.com/vsi